SUPERSPA

THE GOLDEN GATE BRIDGE

50

SUPERSPAN

PHOTOGRAPHY BY
BARON WOLMAN

TEXT BY
TOM HORTON

DESIGN BY
GEORGIA GILLFILLAN
AND
PHIL CARROLL

CHRONICLE BOOKS • SAN FRANCISCO

THE GOLDEN GATE BRIDGE

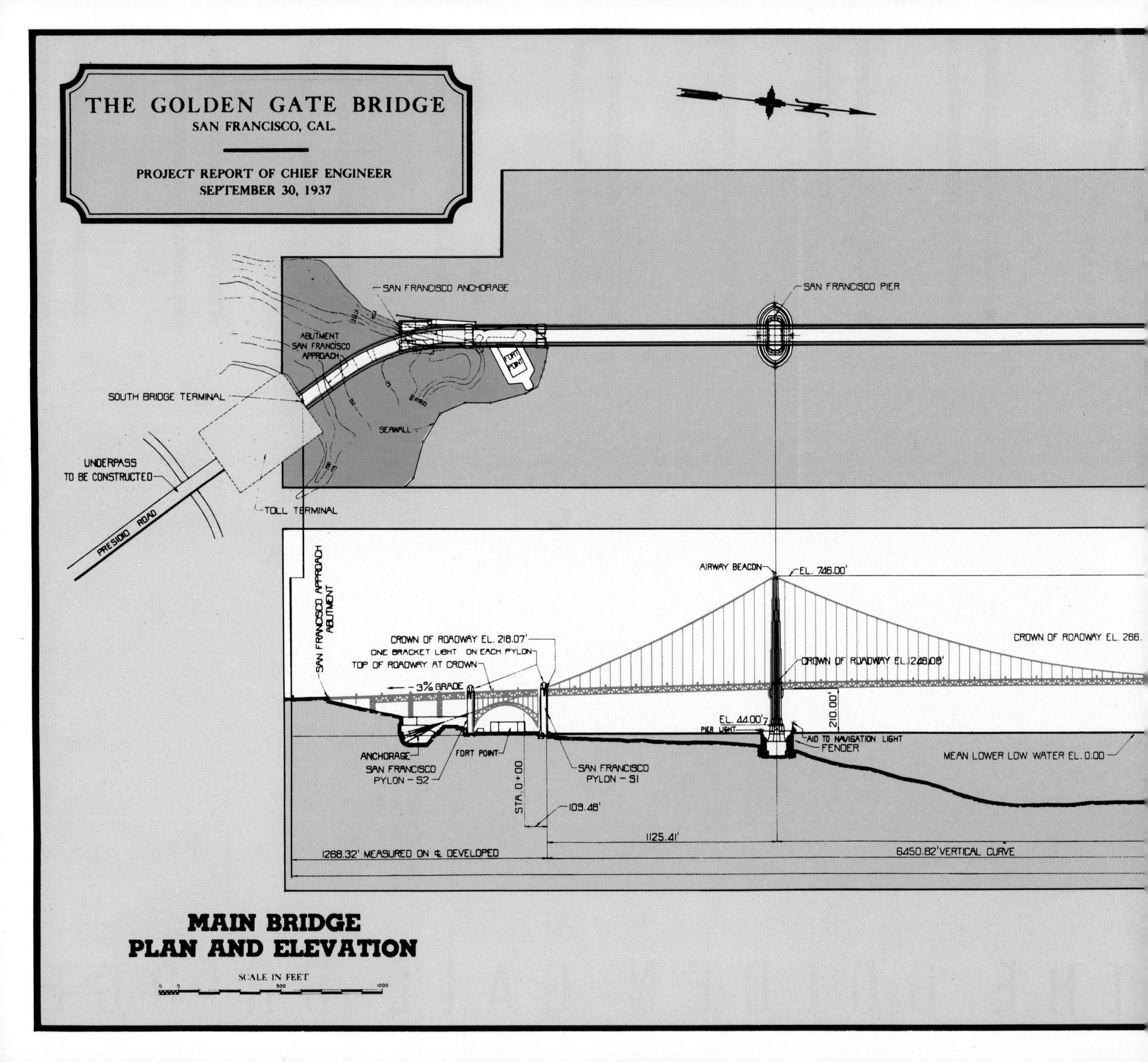
THE GOLDEN GATE BRIDGE
SAN FRANCISCO, CAL.
PROJECT REPORT OF CHIEF ENGINEER
SEPTEMBER 30, 1937
SAN FRANCISCO ANCHORAGE
SAN FRANCISCO PIER
ABUTMENT
SAN FRANCISCO
APPROACH
FORT POINT
SOUTH BRIDGE TERMINAL
SEAWALL
UNDERPASS
TO BE CONSTRUCTED
TOLL TERMINAL
PRESIDIO ROAD
SAN FRANCISCO APPROACH
ABUTMENT
AIRWAY BEACON
EL. 746.00'
CROWN OF ROADWAY EL. 218.07'
ONE BRACKET LIGHT ON EACH PYLON
TOP OF ROADWAY AT CROWN
CROWN OF ROADWAY EL. 266.
CROWN OF ROADWAY EL. 246.08'
- 3% GRADE
210.00'
EL. 44.00'
PIER LIGHT
AID TO NAVIGATION LIGHT
FENDER
MEAN LOWER LOW WATER EL. 0.00
ANCHORAGE
FORT POINT
SAN FRANCISCO
PYLON - S2
STA. 0+00
SAN FRANCISCO
PYLON - S1
109.48'
1125.41'
1268.32' MEASURED ON ℄ DEVELOPED
6450.82' VERTICAL CURVE
MAIN BRIDGE
PLAN AND ELEVATION
SCALE IN FEET
500
1000

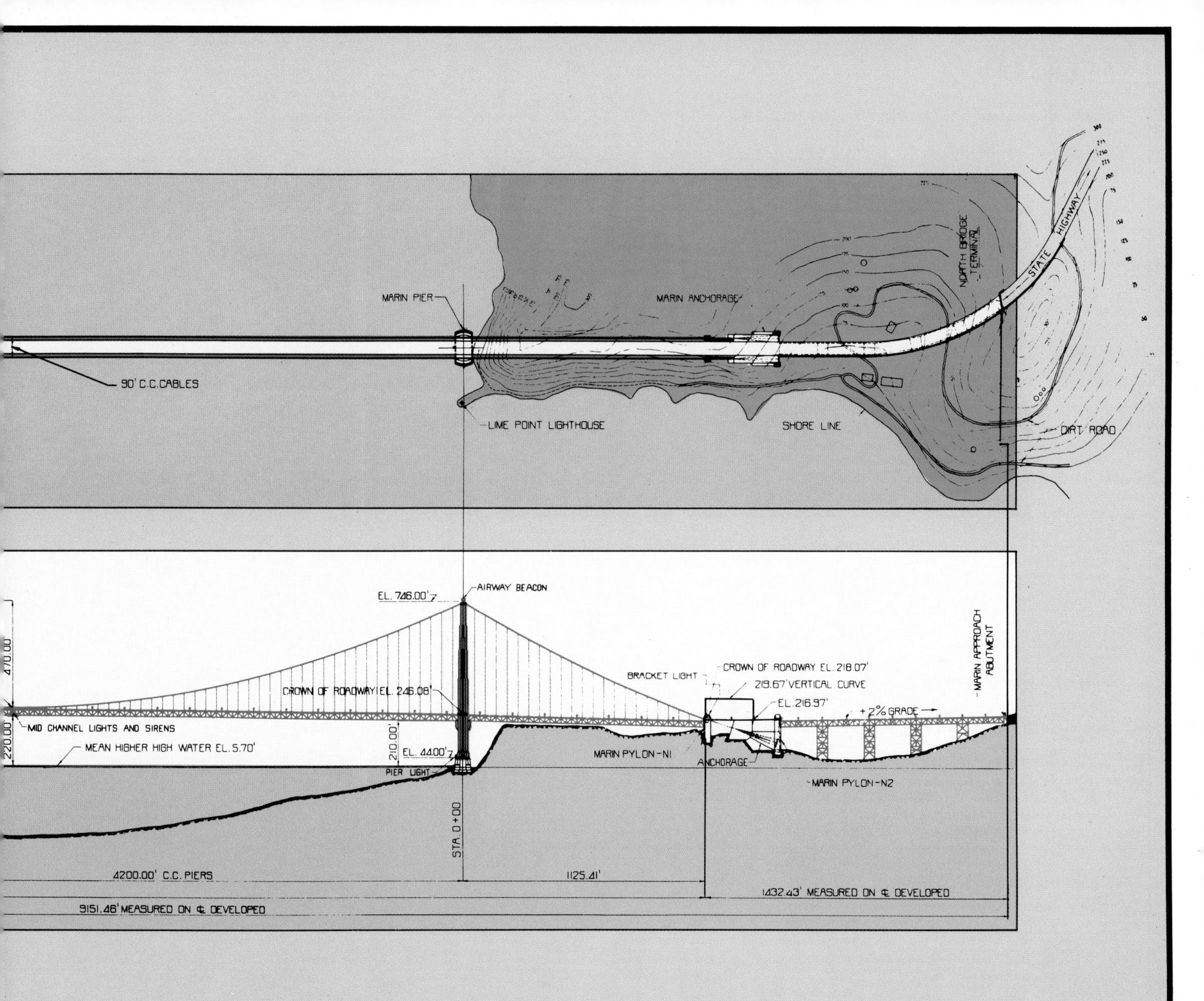
MARIN PIER
MARIN ANCHORAGE
NORTH BRIDGE TERMINAL
STATE HIGHWAY
90' C.C. CABLES
LIME POINT LIGHTHOUSE
SHORE LINE
DIRT ROAD
EL. 746.00'
AIRWAY BEACON
470.00'
CROWN OF ROADWAY EL. 246.08'
BRACKET LIGHT
CROWN OF ROADWAY EL. 218.07'
219.67' VERTICAL CURVE
EL. 216.97'
+2% GRADE
MARIN APPROACH ABUTMENT
220.00'
MID CHANNEL LIGHTS AND SIRENS
MEAN HIGHER HIGH WATER EL. 5.70'
210.00'
EL. 44.00'
PIER LIGHT
MARIN PYLON-N1
ANCHORAGE
MARIN PYLON-N2
STA. 0+00
4200.00' C.C. PIERS
1125.41'
1432.43' MEASURED ON ℄ DEVELOPED
9151.46' MEASURED ON ℄ DEVELOPED

SUPERSPAN

Produced and directed by Baron Wolman/SQUAREBOOKS

Edited by Caroline Sikes

Photo credits: All construction photos are courtesy of the Golden Gate Bridge, Highway and Transportation District. The photos taken at Fort Point on pages 74 and 75 are courtesy of the National Park Service. The photo on page 51, bottom right, of Opening Day, is from the *San Francisco Examiner.* The postcard photo on page 8 is by Charles Weidner of San Francisco, and the postcard itself, as well as the Opening Day pamphlet reproduced on page 52, are from the California Historical Society, San Francisco. Mush Emmons of Sausalito made the photos on pages 68 and 95.

Readers who are interested in a more detailed account of the building of the Golden Gate Bridge are referred to *Spanning The Gate,* by Stephen Cassady, published by Squarebooks in 1979. In 132 pages with text, photos and maps, the history of the most daring and beautiful bridge ever built is told in its entirety. For further information write: Squarebooks, Post Office Box 1000, Mill Valley, CA 94942.

Library of Congress Cataloging in Publication Data

Horton, Tom.
Superspan: the Golden Gate Bridge.

1. Golden Gate Bridge (San Francisco, Calif.) 2. San Francisco (Calif.)—Bridges.
I. Wolman, Baron. II. Title.
TG25.S225H67 1982 388.1'32'0979461 82-17746
ISBN 0-87701-277-6 (pbk.)

Typography by *turnaround,* Berkeley

Printed in Japan by Dai Nippon Printing Co, Ltd

CHRONICLE BOOKS
870 Market Street
San Francisco, CA 94102

SUPERSPAN
IS DEDICATED
TO THOSE WHO
BUILT
MAINTAIN
CROSS
AND CHERISH
THE GOLDEN GATE
BRIDGE

1 AN IMPOSSIBLE DREAM

Juan Rodriguez Cabrillo, Francis Drake, Sebastian Cermeno and Sebastian Vizcaino, skilled explorers all, sailed past it between 1542 and 1602. Fog helped to conceal it. Not until 1769, nearly three centuries after the historic voyage of Columbus, was the Golden Gate seen by anyone except the Coast Indians who had fished and hunted there for centuries. Spanish soldiers found it because they were on foot and lost.

Don Gaspar de Portola left the site of *Mission San Diego de Alcala* on the first recorded overland trip up the California Coast, bound for Monterey Bay. But the Spanish detoured around the formidable Santa Lucia mountains into the flat Salinas Valley, overshot Monterey and became lost on the San Francisco Peninsula. A bewildered de Portola dispatched Sergeant Jose Ortega with a scouting party to find a way out, perhaps by pushing on north to Point Arena, where Drake, Cermeno and Vizcaino had found safe harbor.

Sergeant Ortega led his party up the forested shoreline, past a length of beach where they saw strange outcroppings of rock and heard the barking of sea lions. The riders, continuing along a precipitous rise of cliffs at ocean's edge, suddenly reined their horses to a dead stop and stared in disbelief. Unnerved, the men rode back to camp with the bad news: the way north was blocked by an insurmountable impediment, a bay of untold enormity that began at a treacherously narrow cleavage in the coastline and cut through the earth in all directions—how far, no man would venture to say for none could see an end to it. A dispirited de Portola turned back, found his way into the Santa Clara Valley and eventually located Monterey.

A priest in Ortega's party tried to describe for the Spanish in Monterey what the men had seen that first day of November, 1769: "It is a harbor such that not only the navy of our Catholic Majesty but those of all Europe could take shelter in it."

There was no haste to return. The discovery of San Francisco Bay was of less importance than Captain Vizcaino's discovery in 1602 of *El Puerto de Monterey.* Monterey presented a naturally obliging harbor. The new one to the north was, by contrast, threatening.

It was another six years before the Spanish returned to this vast and unexplored bay. The supply ship *San Carlos* arrived in early August, probably to find the entrance to the bay shrouded in summer fog. Captain Juan Manuel de Ayala anchored outside the inlet, perhaps hoping the fog would lift. Darkness was approaching. Captain de Ayala could wait no longer. Slowly, carefully, the captain steered his little ship through this strait of uncertainty, strong winds testing the mainsails while the wooden hull felt the force of tons of compressed waters making a furious rush to join up with the open sea. A cheer—or more likely a collective sigh of relief—must have been lifted into the chill and fog as the *San Carlos* cleared the channel and on August 5, 1775, became the first ship to enter the bay.

A month later cannons fired from the *San Carlos'* decks as a shore party raised the flag of Spain. A mission would be built here. In honor of the patron saint of the Franciscan order, it would be named *San Francisco de Asís.*

Mission San Francisco was a sorry misadventure, a disgrace to the good name of the Franciscans. Remote, with a less hospitable climate than the missions to the south, it barely survived during a long period of decline and deterioration. Virtually abandoned by the Spanish, the presidio was guarded by barely half a dozen soldiers in the 1820s and the mission was often without a priest. A British sea captain, dropping anchor in 1825, described the presidio as "little better than a heap of rubbish and bones, on which jackals, dogs and vultures were constantly preying."

It was not until 1835, more than half a century since the *San Carlos* had sailed into the bay, that the first evidence of a civilian settlement rose on the shores of this great natural harbor. Ships trading in hide and tallow had begun calling infrequently. Aboard one of them, out of Boston, was a young man of 20 named Richard Henry Dana, who later wrote the classic *Two Years Before the Mast,* describing the view from the deck of the *Alert* as it lay off Yerba Buena Cove in the winter of 1835–36:

Beyond, to the westward of the land-place were dreary sandhills, with little grass to be seen, and few trees, steep and barren, their sides gullied by the rain. Some five or six miles beyond the land-place, to the right, was a ruined Presidio, and some three miles to the left was the Mission of Dolores, as ruinous as the Presidio, almost deserted, with but a few Indians attached to it, and very little property in cattle. Over a region far beyond our sight there were no human habitations, except that an enterprising Yankee, years in advance of his time, had put up, on the rising ground above the landing, a shanty of rough boards, where he carried on a very small retail trade between the hide ships and the Indians.

This shanty was the birth of Yerba Buena, meaning good herb. There were presumably some herbs struggling to take root in this desolate shoreline dominated by two immense sand dunes, but in far greater profusion were wild rabbits and sand fleas. The village of Yerba Buena, as the mission before it, languished in obscurity. If there was anything of significance happening it was mainly of provincial esteem, such as a village celebration in 1844 on the creek at Clarkes Point on the Embarcadero. The celebration was for Yerba Buena's first bridge.

The village grew almost imperceptibly. No more than 30 crude houses huddled between the sand dunes of Yerba Buena in early 1846 when John C. Fremont, captain of the Topographical Engineers of the U.S. Army, led an expedition to California. Fremont later described the dramatic entrance to San Francisco's little-used harbor:

Approaching from the sea, the coast presents a bold outline. On the south, the bordering mountains come down in a narrow ridge of broken hills, terminating in a precipitous point, against which the sea breaks heavily. On the northern side, the mountain presents a bold promontory, rising in a few miles to a height of two or three thousand feet. Between these points is the strait—about one mile broad in the narrowest part, and five miles long from the sea to the bay. Passing through this gate, the bay opens to the right and left, extending in each direction about thirty-five miles. . . . To this gate, I gave the name Chrysopylae, or Golden Gate, for the same reasons that the harbor of Byzantium was named Chrysopylae, or Golden Horn.

So it was named: Golden Gate.

The first civilian settlement inside the Golden Gate, Yerba Buena, had never been more than a bleak, unpromising village. San Francisco had virtually no time to be a village before it became a city—a city of ships and of gold. It leaped from village to city, from obscure harbor to world port of magic and myth, with the speed of lightning. It struck in the middle of the first month of 1848 when the Swiss settler, Johann (John) Augustus Sutter, who had passed through Yerba Buena not ten years earlier, sent the carpenter John Marshall into the foothills above Sutter's Fort in the Sacramento Valley to build a sawmill. Marshall, instead, found gold.

"*Gold! Gold!* Gold from the American River!" the Mormon merchant Sam Brannan shouted at the top of his lungs as he raced his horse through the dusty streets of San Francisco, waving above his head a vial of proof. Just one year earlier the village had officially changed its name from Yerba Buena to San Francisco. At that time, early in 1848, San Francisco was a village of 150 houses and a civilian population of less than 500. Within two years it was a brawling city of 35,000.

The Golden Gate, so quiet for so long, had been given an uncommonly prophetic name, for it was now indeed the portal to gold through which passed the greatest mass movement in the history of the American conti-

The entrance to San Francisco Bay is often shrouded by fog in the summer. Beautiful from above, it is treacherous for seamen. On August 5, 1775, the Spanish supply ship *San Carlos* became the first to clear the channel of the Golden Gate and enter the bay.

nent. In 1848, as word traveled slowly around the Cape, 15 vessels sailed through the Golden Gate. In 1849, 775 came from Eastern ports alone. Yerba Buena Cove sheltered a forest of masts, nearly a thousand ships anchored at one time. San Francisco was a city connected to the world by ships and it became, literally, a city built on ships. Ships, abandoned by crews joining the headlong rush to the gold fields, were hauled to the shoreline and sunk, helping to fill in the bay and provide instant waterfront property for more stores, more hotels, more saloons, more gambling halls and, of course, more bordellos. Ships became landlocked as the bay was filled in around them. Ships were used as warehouses, and as houses in which to store men. Ships became hotels. The city's first prison was the brig *Euphemia.* Public hangings were conducted from the tops of ship derricks. Ships that couldn't be put to use while afloat were scrapped for lumber.

By the time the mail ship *Oregon* sailed through the Golden Gate in 1850, cannons booming to announce California's admission to the Union, San Francisco had become one of the great ports of the world. This overnight fame turned almost as quickly to legend. The teeming San Francisco waterfront, with its infamous Barbary Coast, was soon the equal of any in commerce, and surpassed most in providing for pleasures of the flesh. Italians came to launch fishing fleets, sailing out the Golden Gate and returning with boatloads of salmon, cod, halibut, sole and crab. By the 1860s the Chinese had established a prospering shrimp industry.

True, San Francisco was a peninsula, cut off by the bay from the fertile, open, sunnier lands to the north and the east. But as early as 1826 there was a sloop ferrying passengers across the bay. Two 30-ton passenger schooners plied the bay as Yerba Buena took shape in 1835. Soon after the Gold Rush began, the bay accommodated its first steamboat, the 37-foot sidewheeler *Sitka,* brought down in pieces from the Russian settlement in Alaska. The *Sitka* could take miners from San Francisco upriver to Sacramento in only six days. By 1850 the propeller steamer *Kangaroo* inaugurated ferry service between San Francisco and Oakland. In 1863 the *Contra Costa* began crossing six times daily. The *Sophie McLane* introduced train-ferry service from Alameda to San Francisco and in 1869 the steamer *Alameda* welcomed aboard the first San Franciso-bound passengers to reach the Pacific Coast on the transcontinental railroad. By 1879 whole trains were being floated across the bay on the world's largest ferryboat, the *Solano.* True, San Francisco was a peninsula, but access to the city from any direction was not a problem.

In such a dynamic scene of

maritime prosperity, of noble clipper ships and brawny steamers, it would have taken a madman to suggest that this great harbor was missing anything. Joshua A. Norton fitted the bill perfectly. He had good reason to hate ships. And in San Francisco's legendary galaxy of eccentrics, he was the shining star.

An Englishman, Norton had arrived in San Francisco one year after the gold discovery, made a fortune, then lost it, along with a good share of his mind. He sank from sight, only to resurface years later, but no longer behaving as a respected member of the business community. In 1859 the *San Francisco Bulletin* printed Norton's announcement that he was, "At the peremptory request and desire of a large majority of the citizens of these United States," Emperor. Later, for good measure, he added, "Protector of Mexico." He dressed the part: long, dark coat cut military style, swarming with brass buttons, and shoulders bearing oversized gilt epaulets. San Francisco's tolerance, even encouragement, of unconventional behavior—so long as it was entertaining—was already well ingrained in the city's character. Emperor Norton he would be, and the city gleefully submitted to his rule, accepting his hilarious announcements, described by him as "proclamations," with mock solemnity, affording him the privileges of royalty (free meals, opening-night theater seats), raising him to heights of nonsensical nobility never attained by any of Emperor Norton's formidable compatriots, not even by Oofty Goofty, who for four bits would let you hit him on the head with a baseball bat.

Emperor Norton's most famous proclamation was, perhaps, no more than an exceptional extension of his imbalanced mind into the far reaches of madness. Or maybe he had never forgotten how he came to ruin in 1853. An already wealthy Norton had attempted to corner the rice market and increase his riches by buying up all the rice on the West Coast and holding it until the price soared. But as he anticipated his coup, three ships sailed unexpectedly through the Golden Gate, loaded with rice. The bottom dropped out of the rice market, and with it fell Joshua Norton. Is it possible that for the rest of his life he hated the sight of cargo ships? That whatever was left of his mind saw each ship sailing through the Golden Gate as the one bearing the seeds of his financial ruin, and that he favored the notion of any alternative to reliance on capricious maritime transport? No one knows. But we know that he delivered of his royal mind a memorable proclamation.

The city howled. Imagine, a *suspension bridge* across the Golden Gate. What amusing madness.

PROCLAMATION

WHEREAS, it is our pleasure to acquiesce in all means of civilization and population:

Now, therefore, we, Norton I, *Dei Gratia* Emperor of the United States and Protector of Mexico, do order and direct first, that Oakland shall be the coast termination of the Central Pacific Railroad; secondly, that a suspension bridge be constructed from the improvements lately ordered by our royal decree at Oakland Point to Yerba Buena, from thence to the mountain range of Saucilleto and from thence to the Farallones, to be of sufficient strength and size for a railroad; and thirdly, the Central Pacific Railroad Company are charged with the carrying out of this work, for purposes that will hereafter appear.

Whereof fail not under pain of death.

Given under our hand
this 18th day of August, A.D. 1869.

NORTON, Emperor The First

THE DREAM

James Wilkins traveled to work on a ferryboat. Everyone who lived in Marin County and worked in San Francisco in 1916 went to work on a ferryboat. It was inconceivable to most ferryboat riders that anyone would ever get from Marin to San Francisco any other way. Occasionally there would be talk of building a bridge across the Golden Gate. But few people took such talk seriously, not even in 1872 when it was proposed by Charles Crocker, who had personally supervised construction of the transcontinental railroad across the awesome Sierra Nevada.

Many people loved the ferryboats. James Wilkins was not among them. He was a newspaper man, impatient, accustomed to deadlnes, intolerant of delayed action. And like all crusading newspaper men of the day, he was a champion of progress. The rest of the country was moving forward on wheels. In California alone there were half a million automobiles. In an automobile a man could go 20 miles in half an hour, maybe faster if the road was good. It took Wilkins and all the other daily ferryboat passengers longer than that just to go barely five miles across the bay to get to work. Some 100,000 automobiles a year were being carried from Marin to San Francisco on the ponderous ferryboats.

Some didn't mind the trip. A ride across the bay was not merely a way to get to work; it was a social occasion, a twice-daily party of friends, neighbors and co-workers; it was a daily reminder that they had the best of two worlds, a job in the most beautiful, vibrant city in the world, and a home in the quiet countryside of Marin.

There was none of the ferryboat romanticist in James Wilkins. He had graduated from the University of California at Berkeley with an engineering degree, although he later chose newspaper work. He had been raised in Marin, started a small paper in Marin, and he preferred to continue living in wooded Marin County (where he would later become mayor of San Rafael). But in 1916 Wilkins worked for one of the papers in the city, the *San Francisco Bulletin.* That put him on the ferryboat twice a day and it galled him. He was not so entranced with the bucolic value of Marin that he couldn't become excited about what a bridge could do for commerce and development on the north side of the bay. Wilkins treated his isolation from San Francisco as more than a personal nuisance. He saw it as a major impediment to the progress of his beloved Marin County.

The August 26, 1916 edition of the *Bulletin* carried a bylined article by Wilkins under the headline, "Suspension Bridge for Golden Gate Would be Biggest in World." Wilkins wrote:

It is possible to bridge San Francisco Bay at various points. But at only one point can such an enterprise be of universal advantage—at the water gap, the Golden Gate, giving a continuous dry-shod passage around the entire circuit of our inland sea.

The northern counties, almost an empire in extent, with potentialities barely surface-scratched, contain a present population that has passed well beyond the 200,000 mark. Nature has, in a way, tied their fortunes, beyond recall, to San Francisco. That city must always be their final marketplace —the clearing house. Nothing can be more important to San Francisco, beset as it is by jealous rivals carelessly endeavoring to win away its trade, than the speedy development of a region whose business comes to it automatically, which can never be diverted elsewhere. And nothing can hasten that development more than the free circulation of modern life by a bridge across the Golden Gate.

Thus did one ferryboat commuter from Marin launch a vigorous campaign to span the Golden Gate. He used his engineering background to tell how it could be done, estimating the cost at $10 million, and he played upon San Francisco's

The 6,450 foot long Golden Gate Bridge dwarfs the 102 foot long Golden Hinde II as she sails into San Francisco Bay in March, 1975. Crowds of well-wishers hail greetings from the east walkway.

THE GOLDEN HINDE

Sir Francis Drake was premier adventurer. The first Englishman to sail around the world, he was both courageous and aggressive. Before he was 30 years old—and with the encouragement of Queen Elizabeth I—he had methodically loosened $48 million worth of bullion, precious stones and jewelry from the holds of Spanish ships.

In 1577, Drake and four other ships departed Plymouth, England. Before the vessels reached the Straits of Magellan, only three were left. In a monster storm near the Straits two more were lost. Drake alone continued up the western coast of South America, a one-ship raiding party. In June 1579, after weeks of looting and plundering on land and sea, he sailed his barque—the Golden Hinde—into San Francisco Bay for a few days of rest and relaxation.

Nearly 400 years later, in March of 1975, another Golden Hinde returned to San Francisco Bay after a five-month, 13,000-mile journey from Plymouth. Allegedly a replica of Drake's vessel, the Golden Hinde II was a square-masted warship, only 102 feet long and 20 feet wide. Considerable effort had been made to maintain authenticity during its design and construction. Since there were no blueprints of the original Golden Hinde, a naval architect from Marin County spent three years of research on the project. The finished ship was "built along Venetian lines and represents the transition from the carrack to the galleon."

This latest voyage from England to California had been tame compared to Drake's. While Drake was forced to sail south around Cape Horn and through the Straits of Magellan, the Hinde II enjoyed the luxury of passage through the Panama Canal. The crew of the Hinde II numbered 18 men, each with a bunk of his own. Drake had 80 men on board, many of whom slept on the deck. Only 10 miles out of San Francisco, the Hinde II had encountered severe weather. Pounded by gale winds of up to 60 knots and swells that reached 35 feet, the barque was tossed around for hours but suffered no serious damage. The furious storm near the Straits that destroyed Drake's two sister ships lasted an unbelievable 52 days.

Such differences were of little interest to San Franciscans. The arrival of the Golden Hinde II was simply a cause for celebration. The little ship was greeted by a flotilla of pleasure boats, Coast Guard cutters and fireboats. Thousands of spectators lined the shore and stood on the Golden Gate Bridge to wave their "hellos."

The replica of Sir Francis Drake's tiny barque was escorted into the harbor by hundreds of private sailboats, coast guard cutters, fireboats, helicopters and small planes.

love for the grandiose to tell how it would do the city proud. He pointed to other heroic forms raised by man, from the Colossus of Rhodes and the Pharaohs of Alexandria to the Statue of Liberty. "But the bridge across the Golden Gate," trumpeted Wilkins, "would dwarf and overshadow them all."

Wilkins's crusade naturally met with the approval of those tired of loading their automobiles onto ferryboats, but more importantly it struck a responsive chord with city engineer Michael Maurice O'Shaugnessy. O'Shaugnessy relished a challenge. Two years earlier he had proposed a grand idea for conquering the Fillmore Street Hill: a movable sidewalk. On a larger scale, he had masterminded the massive $100 million Hetch-Hetchy project that brought fresh water from the Yosemite Valley, 156 miles distant, to San Francisco. He had cut a streetcar tunnel through Twin Peaks and engineered the city's first automobile tube, the downtown Stockton tunnel. To Michael O'Shaugnessy, building a bridge across the Golden Gate sounded like a smashing idea. His endorsement was critical, since he was the first city official and the first practicing engineer to publicly support a Golden Gate Bridge.

The bridge was vigorously debated. Even if it was possible to build a bridge across one of the most treacherous channels in the world, and few men with engineering credentials believed it was possible, many critics argued that the cost would be prohibitive. Dismissing the newspaper man's estimate of $10 million, some skeptics charged that even if it could be built, it might cost as much as $100 million. Such a figure indeed sounded prohibitive, for this was to be an undertaking of cities and counties, not of state or federal governments. There was no hint of state or federal government interest in financing a bridge—and a questionable bridge at that—to benefit the merchants and motorists of the Bay Area. If the citizens on either side of the bay wanted a bridge, they would have to build it themselves.

After the initial flurry of excitement, the bridge issue faded once again into the background. A larger challenge was at hand, the war in Europe. Bridge lobbyists waited. After the armistice, James Wilkins, still riding those ferryboats every day, fired up his bridge campaign anew. Americans had just won a world war and they were more cocksure of Yankee ability to prevail against the odds than ever before. And no city was more sure of itself than San Francisco. Wasn't it only a dozen years ago that they'd rebuilt the whole city from the ashes and rubble of earthquake and fire? Why *not* a bridge across the Golden Gate? The San Francisco Board of Supervisors, sensitive to public sentiment, moved boldly forward in late 1919 by directing engineer O'Shaugnessy to begin an official study. The courage demonstrated by the supervisors was somewhat diluted by the amount of money they appropriated for the study: zero.

O'Shaugnessy invited three highly regarded bridge engineers to submit detailed designs and cost estimates for a Golden Gate Bridge. One of them was a bridge builder headquartered in Chicago named Joseph Baermann Strauss.

Joseph Strauss was not a likable man. Those closest to him described him privately as rude, pompous, vain, short-tempered, obstinate, impatient, even a bully; also, brilliant, courageous, tireless, inspired, indomitable, a romantic, an artist, even a poet. He was all of these, and how he was judged by his contemporaries depended upon whether they were judging the man or his works, and of course whether or not they agreed with his vision. Strauss didn't care what they thought, if only they would step aside and let him build his bridges. He once described his success as "hard work and an absolute unwillingness to be discouraged by what others say."

He was born in 1863 in Cincinnati and it has been suggested that his future was decided the first time he contemplated the shape of the 1,057-foot main span of the Cincinnati Bridge, proclaimed the "Biggest Bridge in the World" when it crossed the Ohio River into Kentucky in 1866. Legend, and it is probably just that, has a young Strauss staring resolutely at the bridge from the window of the University of Cincinnati infirmary after nearly being torn to pieces in a tryout for the football squad. The pint-sized Strauss, who as an adult barely topped five feet and weighed about 120 pounds, is supposed to have vowed from his sickbed that he would raise himself to vicarious heights of glory by erecting heroic bridges. Hollywood couldn't have invented a more dramatic scene for *The Life of Joseph Strauss, Master Bridge Builder, Act I.*

During his sophomore year Strauss changed his major to engineering, and revealed a natural talent for mathematics. But he was not one to be content with the quiet pursuit of technical expertise. He was instinctively artistic as well, and his multiple talents, combined with an ego craving recognition, were sufficient to create the passions of a revolutionary. He did not see his destiny as improving upon what was being done in his field, but as doing what had never been done before. To the practical tenets that bound nineteenth-century engineering he brought dreams that fired the imagination and recognized no boundaries. Years later, trying to sell one of those dreams on the Western rim of the continent, he would make the unthinkable sound almost inevitable: "Our world of today revolves around things which at one time couldn't be done because they were supposedly beyond the limits of human endeavor. Don't be afraid to dream."

A clue to where Joseph Strauss's dreams would lead him came in his graduation thesis in 1892. It was the design for a bridge—a bridge 50 miles long, across the Bering Straits, from North America to Asia.

By 1919 Joseph Strauss, 49 years old, had a solid reputation has one of the world's great bridge builders. He had built more than 400 bridges, not only throughout the United States and Canada, but in the mountains of Panama and even across the Neva River at the Winter Palace of the Czar of Russia. He had revolutionized the movable bridge industry in the early 1900s by engineering a new style of bascule bridge, using concrete instead of expensive iron for the mechanized counterweights that raised and dropped the bascule bridge span in what was a modern advancement of the ancient drawbridge. For the next dozen years Strauss would continue to introduce revolutionary concepts, continue to build bridges at a prolific rate, always in search of the challenge that would take him to new heights.

The timing was perfect. The stage was heavensent. A Golden Gate that couldn't be spanned. A bridge builder who wouldn't be denied.

As San Francisco danced merrily into the 1920s with the rest of the country, the bridge was finally becoming a subject of serious debate. It was a time of unbounded optimism among Americans. Opposition to the impossible became opposition to the possibly unaffordable. The first estimate from a famous Eastern engineer put the cost of a Golden Gate Bridge at a staggering *minimum* of $56 million. Then Strauss, unaware that two other engineers had been asked to submit plans, delivered his. The Strauss estimate: $17 million.

The Strauss estimate was a welcome surprise. The Strauss design was a shock. He proposed to build a "symmetrical cantilever suspension bridge," his own innovative marriage of the suspension and cantilever methods. Explained Strauss, "The cantilever by itself would have too large a self-weight for a span across the Gate, and so long a suspension bridge would lack the necessary rigidity and involve huge costs. The combinations of the types uses the best features of each. . . ."

One feature which the design most certainly did not possess was beauty. A ponderous mass of brute steel, it was jarringly alien to the great natural beauty of the land and sea.

This unorthodox design did not inspire the immediate public outcry that might have been

Twice a day fresh water from the California mountains and rivers rushes to sea through the Golden Gate; twice a day water from Pacific returns to the Bay. These strong tides regularly force 2,300,000 cubic feet of water per second through the narrow channel, at speeds between 4.5 and 7.5 knots per hour.

anticipated. The proposal was, according to some of the country's leading engineers, a sound one, and perhaps a skeptical public wanted to know whether *any* bridge could be built across the Gate before concerning itself with aesthetics.

The great bridge debate, smoldering for so long, finally caught fire. Strauss, despite a well-deserved reputation as a boring speaker, took his vision to the most naturally receptive audience—the 200,000 residents of the North Bay counties who stood to benefit from a span connecting them and their automobiles and delivery trucks to San Francisco. Momentum built. On January 14, 1923, powers from both sides of the bay and perhaps a hundred interested taxpayers met at the Santa Rosa City Hall in Sonoma County. Strauss was unable to attend, but O'Shaugnessy, San Francisco Mayor James Rolph and other local officials were present. From this small but influential group came the first politically motivated body whose purpose was clearly defined by its name: Bridging the Gate Association. Joseph Strauss was named engineering consultant, without salary.

Barely four months later, Governor Fred W. Richardson signed a bill granting voters the power to form a tax district. An elated Strauss sounded ready to begin pouring cement: "The Golden Gate will be bridged and opened for traffic by 1927," he proclaimed, adding two minor ifs: ". . .if the people of San Francisco and other communities of the Bay region are willing to spend $20 million and if they are successful in obtaining the sanction of the War Department."

The money might prove less of an obstacle than the War Department, which had jurisdiction over anything that might interfere with ship traffic in San Franciso Bay. For years the military had exhibited a near-paranoia on the subject of bridges in San Francisco Bay. San Francisco school children were assigned by their teachers to write essays on why a Golden Gate Bridge shouldn't be built because the Japanese might bomb it and block the harbor. The War Department had killed proposals for a San Francisco–Oakland Bay Bridge in 1915 and again in 1921. Now the War Department held a public hearing to consider the dangers of a bridge across the Golden Gate. The U.S. Army district engineer said the decision would turn on two points: "The first and most important is whether or not the bridge will constitute a menace and a hindrance to navigation both in peace and in war. The second is whether or not it will be adequately financed."

Strauss responded with typical hyperbole: "San Francisco has often done the impossible. Now it only remains for her to connect up with the contiguous territory to make her the great city she is destined to be. Ways of transportation are essential to a city's welfare; it means the decline or growth of a city. I believe this bridge will bring an era of unprecedented prosperity. It will be, in my opinion, the greatest feat of construction ever developed."

No matter that San Franciscans, and a good many people of good taste in other cities of the world, already considered San Francisco a great city and had so considered it for more than half a century. Or that it was the North Bay counties whose welfare depended on spanning the Gate. To Strauss you couldn't have a great city on a great harbor without a great bridge.

The War Department was not concerned with man's building genius, but with his destructive powers. If the bridge were bombed, would the harbor be blocked? Perhaps, said Strauss, but the harbor could just as quickly be blasted open. Unable to resist a sarcastic commentary on military thinking, he added, "If the enemy got so close as to be able to bomb the bridge, there would be very little left of the city." His analysis proved timeless. In 1982 Bay Area politicians would be arguing over how the bridge should be used to evacuate San Francisco after a nuclear attack.

It was near the end of 1924 before Secretary of War John Weeks announced that the War

Department would withdraw its opposition to the bridge if certain conditions were met. Essentially, the conditions were that the bridge district pay all costs for moving any military installations flanking the Gate, that the U.S. have control of the bridge in time of war, and that all government vehicles ride free across the bridge. The conditions were accepted.

Bridge supporters were jubilant. A major hurdle had been cleared and the preliminary work could begin. It soon became clear that this optimism was premature.

The world's largest suspension bridge would also be a remarkable project in terms of financing. No state or federal funds would be used. And while it might have seemed to the casual observer that this was a San Francisco bridge, its fate rested beyond the city, in the pockets of taxpayers scattered along the North Coast, from Marin on the shores of the Golden Gate to tiny Del Norte County 350 miles distant, which shared its northernmost border with Oregon.

Political bickering swept the North Coast with the intensity of a winter storm. Counties began withdrawing from the bridge district at an alarming rate. First Humboldt County, then Lake County defected. Mendocino County, first to join in 1925, changed its mind and voted to withdraw a year later. This left five counties: San Francisco, Marin, Sonoma, Napa, and, still faithful, Del Norte. Individual property owners were allowed to protest inclusion in the district, and the protests poured in. A judge was appointed and a lengthy series of hearings began. The bridge was attacked from all sides. Experts testified again that it was unsafe, that it couldn't be built, and again, that even if it could it would cost five times what Strauss was claiming. The Joint Council of Engineers submitted an independent study which they said proved that Strauss's design was recklessly dangerous. They scoffed even at his refigured estimate, now $21 million, claiming that it would cost closer to $112 million. The higher estimate, if accepted by the court, would be enough to bury the bridge dream once again.

In 1928 Christmas came on December 1 for those who believed in the bridge. Judge C. J. Luttrell, who had conducted six months of hearings through five counties, then spent an additional seven months studying the arguments of both sides, ruled in favor of the bridge district. "The cost of construction will not be prohibitive as compared with the revenues reasonably to be expected from the operation of the bridge. The project is feasible both from the standpoint of an engineering and a financial undertaking," said the judge. The verdict was not unanimous. Some 80 percent of Napa County and 24 percent of Mendocino County were allowed to withdraw from the district. But San Francisco, Marin and Sonoma remained intact, as did Del Norte, the little neighbor to Oregon that had not made a single official protest to joining the Golden Gate Bridge and Highway District.

Again, bridge promoters celebrated the end of political debate, triumphant in their certainty that the work would soon begin. Again, their confidence was premature. One squabble followed another, controversy fed on itself, and a pattern of petty politics that would cling to the bridge forever established itself early. There was a dispute over the appointment of the first bridge district directors from San Francisco. Critics of Strauss refused to yield. There was even doubt that he would be the man to build the bridge. Strauss was only one of 11 candidates who were seriously considered for the position of chief engineer. To stay in the running for the job he had now pursued for nearly a decade, he had to turn politician himself—no easy role for a man accustomed to dealing brusquely with people. In August of 1929, after months of reviewing the candidates and of indulging in the obligatory backroom politicking, the bridge directors named Joseph Strauss chief engineer for the construction of the Golden Gate Bridge.

In the midst of the Great Depression, Strauss finally was ready to call for construction bids. While the free-spending 1920s would have been the ideal time to finance the Golden Gate Bridge, the 1930s should have been the worst of times. In a sense, the Depression actually worked to the bridge builders' benefit. Voters were asked to accept $35 million in bonded indebtedness at a time when 14 million Americans were out of work. But unemployment became the campaign battlecry for building the bridge that had been put off during better times. "It's the job of every voter in the city," declared the president of the San Francisco Chamber of Commerce, "to create jobs by voting for the bond issue." Jobless voters agreed. The bond issue carried by three-to-one.

Surely, the bridge promoters exclaimed, the last conceivable barrier had been removed. But it hadn't. When Golden Gate Bridge bonds were offered, there were no takers. New qualms were stacked on top of old ones. Even the bridge's first public supporter, city engineer O'Shaugnessy, withdrew his support, saying it was a good project at the wrong time. Then the bridge district's constitutional right to tax, and therefore to guarantee the bonds, was challenged. Yet another court test had to be endured. Southern Pacific Railroad, the rich and powerful trust that had ruled California with an arrogant hand for half a century, was accused of conspiring to block the bridge. The charge was not without foundation. Southern Pacific owned a monopoly on the ferryboats and by 1930 it was the largest transportation system of its kind in the world, carrying more than 40 million passengers across the bay in all directions. A court ruling did not come until July of 1932, a decision clearly in favor of the bridge district. But Southern Pacific-Golden Gate Ferries Ltd. announced that it would appeal all the way to the U.S. Supreme Court if necessary, meaning untold delays—and costs. A project thus delayed could easily become a dead project. But public pressure on Southern Pacific forced the lords of the rails and the ferries to surrender. Damning the bridge to the end, Southern Pacific withdrew its legal challenge.

The struggle in the courts and in the public halls of debate had ended, so the work could begin. But not until yet another serious problem was solved. The Depression had deepened and the market for the bridge bonds was virtually nonexistent. A financial miracle was needed and it arrived in the person of Amadeus P. Giannini. The Italian immigrant was at his best during a crisis. He had founded the Bank of Italy in 1904 and after the 1906 earthquake he saved the bank from the advancing fire by hauling assets and records out of the city in horse-drawn wagons, camouflaging the valuable cargo under fruits and vegetables. The Bank of Italy was the first to reopen after the earthquake. Giannini helped to finance the rebuilding of San Francisco and in the process he built the Bank of Italy into one of the world's largest financial corporations, the Bank of America. During the Depression he was once again lending money to put San Francisco back on its feet. Like Joseph Strauss, A. P. Giannini was a builder, a man who believed in the American dream.

"San Francisco needs that bridge," Giannini announced. "We will take the bonds."

So the time had come at last. It had been the demented vision of a madman when Emperor Norton first commanded that it be done. It was swept aside when Charles Crocker proposed to try. It was still beyond the grasp of the average person when the ferryboat commuter James Wilkins began campaigning for it in the *Bulletin.* And it had exhausted the faith of its believers to the point of despair as time after time the dream led them to the brink of victory, only to turn them back again. But they had not despaired. And now a new spirit of unity and purpose arose from the city and the hills beside the bay as the faithful set out to prove to themselves and to the rest of the world that yes, it was a dream, but no longer an impossible one.

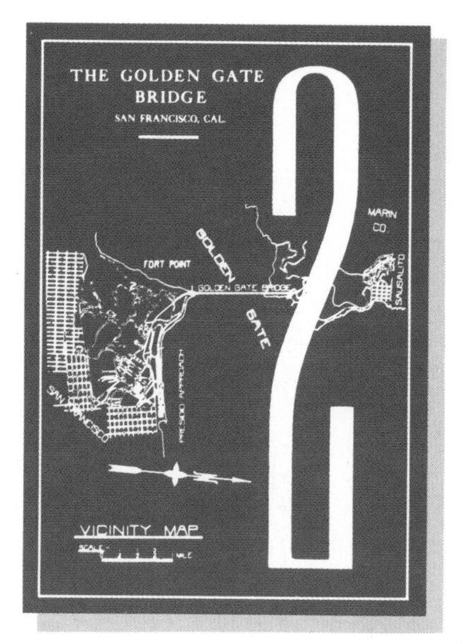

BUILDING THE BRIDGE

It is the perfect place for a bridge. It is also one of the worst places to build a bridge. The Golden Gate averages two miles in width between the shorelines of San Francisco and Marin. The channel's depth ranges from 40 to 300 feet. Winds and water currents are murderously unpredictable. Gusts through the Gate reach 40 miles per hour with distressing regularity, and can turn to gales of 50 to 70 miles per hour during winter storms. The gate is swept by rain in the winter and fogbound most of the summer. It can be wet, cold and inhospitable at any time of the year.

San Francisco Bay is the largest landlocked harbor in the world, with 450 square miles of water. It holds runoff from a vast system of rivers and streams which originate 200 miles to the east in the melting snows of the High Sierra and 400 miles to the north on the upper slopes of Mount Shasta. It serves as a giant holding tank for one of the world's largest natural drainage systems, receiving the runoff from 59,000 square miles of valley, foothill and mountain terrain—nearly 40 percent of the land area of California.

There is but one source of relief for these trillions of gallons of water gorging San Francisco Bay: the open sea. And there is but one route to the sea: the Golden Gate. Like armies in frenzied retreat through a thin mountain pass, billions of gallons of water regularly rush out the Gate at tremendous speeds. Never had a bridge been built across such an intimidating confluence of water. It would have to be the longest and tallest single-span suspension bridge in the world.

By the 1930s there were three famous suspension bridges in America: the Brooklyn Bridge, 1,596 feet of clear span; the Camden Bridge in Philadelphia (1926), a 1,750-foot clear span; and the George Washington Bridge in New York (1931), a 3,500-foot clear span. The Golden Gate Bridge would require an unprecedented center span of 4,200 feet.

It was first designed, in keeping with the taste of the time, as an extravagantly ornate bridge. The masonry toll plazas would be copies of the *Arc de Triomphe,* started by Napoleon and just then nearing completion in Paris. Gates of wrought iron would greet traffic with the emblazoned inscription "Golden Gate Bridge." A glass elevator would carry sightseers to an observation deck at the top of the south tower. Joseph Strauss would build a bridge that was not only the engineering marvel of the world, but as florid as Stauss's poetry, as flamboyant as the man himself.

Good fortune interceded. Perhaps the Golden Gate was inordinately blessed by the gods of check and balance. Fate had been kind enough to send the right man to build the bridge, and, of preemptive importance, to sell the very idea of a bridge. Now the fates kindly intervened with moderating influences on the precocious engineer who meant to prove himself the Picasso of bridge-builders. Most of the ornate features of the Strauss design would ultimately be eliminated. The changes were not made in a single stroke of genius. There was a gradual accumulation of reversals in direction. Budget considerations helped to remove some of the unnecessary decorations. Two men, Clifford Paine and Irving Morrow, were instrumental in reshaping Strauss's original vision.

Luck had always played a role in the evolution of the bridge that would span the Golden Gate. If the ferryboats had not begun to carry railroad trains across the bay, and if Charles Crocker had been more persistent, the first bridge across the Golden Gate might have been a black and brawny railroad trestle. If the San Francisco harbor had not been so strategic to West Coast naval defense and the War Department not so fearful of enemy bombers, a bridge might have been built sooner, and it would have been a different Golden Gate Bridge. If all the forces aligned against the bridge had not fought it off for so long, Strauss might have kept his promise in 1923 that "the Golden Gate will be bridged and open for traffic by 1927." And Strauss might have built the bridge he designed in 1921, his "symmetrical cantilever suspension" bridge, with railroad tracks running through the center of the span—a bridge portrayed by one critic as "two grotesque steel beetles crawling out from either shore."

Time and circumstance turned the tide in new directions. A bridge delayed became a bridge improved. Progress, more than protest, killed Strauss's cantilever-suspension hybrid. During the years of political debate and stalling tactics, bridge-building materials and techniques advanced sufficiently to convince Strauss that a clear-span suspension bridge, without his cantilever compromise, was now possible. New tests had also proved a suspension bridge capable of withstanding the winds of the Golden Gate. And there was the arrival on the scene of Clifford Paine.

Strauss was nearly 60 by the time he was able to begin building the bridge. To assist him and to oversee the project he called on one of the most brilliant engineers then employed by Strauss's Chicago firm. It was Clifford Paine who was there day by day, to guide the bridge to completion. History would record Joseph Strauss as the man who built the Golden Gate Bridge. The men who did the work would dismiss popular history and point to chief assistant engineer Clifford Paine as the man who actually built the bridge for his boss, Joseph Strauss.

Construction started January 5, 1933, when two steam shovels took the first bites out of Lime Point, where a mammoth pit would be dug for the Marin anchorage. The celebration followed nearly two months later, on February 26, when the official ground-breaking ceremony was held on the San Francisco side of the bridge. The motivation for spanning the Gate may have been to open up Marin and other north counties, but from the beginning the Golden Gate Bridge was a San Francisco bridge. It was San Francisco Mayor Angelo Rossi who shared the handle of the golden shovel with bridge district president William Filmer, to turn the first ceremonial soil. There was a parade through the Marina District to Crissy Field, within sight of what would be the San Francisco anchorage. A large crowd gathered. Governor James Rolph, a former mayor of San Francisco, led the list of pontificating politicians (President Herbert Hoover sent a telegram), Navy planes soared above the bay, and it was a jubilant beginning to what would be four years of progressive high spirits during one of the most depressed eras in American history. For the rest of their lives the builders of the bridge would be unable to talk about it without reminding their listeners that it had all been done during the height of the Depression. Said one of them, "It seemed like the Golden Gate Bridge was the only good thing happening in the whole country."

ANCHORAGES AND PIERS

The first step in construction was a massive one, but not visually exciting. Anchorage blocks for the cables were being built on dry, solid land, one at Lime Point on the Marin side, the other just behind Fort Winfield Scott (now called Fort Point) on the San Francisco side. For each anchorage two pits were dug, each deep enough to hold a 12-story building. Rivers of concrete flowed into the pits and hardened into solid blocks, each weighing 50,000 tons. Between the two anchorage blocks and the four pylons, two at each end, 182,000 cubic yards of concrete were poured. Each anchorage had to resist up to 63 million pounds of pull from the bridge cables.

Work went smoothly. No time was wasted. If bridge workers needed more than normal motivation, they could look to the fence that bordered the job site. Behind it crowds of unemployed men waited and hoped. If a man should quit or be fired, a foreman could walk to the fence and choose a replacement from among many eager, often desperate, men. Barrett & Hilp, which had the $1,859,855 contract for the anchorage blocks

The north pier was close to shore; construction was relatively straightforward. However, the south pier, more than 1100 feet offshore, was sunk 100 feet below the surface in the open sea, amid the powerful tides and surf.

and other smaller concrete structures, could count on a full day's pay ($4 a day for unskilled workers, up to $11 for the specially skilled). Bridge workers had a saying: "Eight for eight or out the gate."

As construction of anchorages and pylons proceeded, the Pacific Bridge Company was preparing for a more formidable challenge: the piers that would hold the highest bridge towers in the world. The contract was for $2,935,000. The Marin pier could be built close to the shore, at the base of a cliff only about 20 feet below the water line. Before they could excavate the pier site, bridge workers had to build a cofferdam, a boxlike structure which would prevent the ocean from flowing into the excavation area. Braced by timbers and filled with broken rock which formed a wall, the cofferdam enclosed three sides of an area 178 by 264 feet; the fourth side was formed by the cliff. After the excavation was pumped dry of water—and fish, which workers happily hauled home by the sackful—dynamite and jackhammers removed another 15 feet of solid rock. A total of 45,000 tons of concrete was then chuted down an "elephant trunk" into forms built at the bottom of the excavation.

Frenchy Gales, who worked on the Marin pier, told what it was like when the concrete began pouring into a hole larger than a football stadium:

A regular river of cement could come down there. Those trucks would dump into a big chute that pointed in the forms, and the two guys working the chute really had a job. They had to be careful that the bottom of the chute didn't hit right where they were pouring. If they had a mound there, and they hit it with new cement, the cement would back up over the top like water from a busted main. I got buried that way once. Oh Jesus, yeah.

I was down at the bottom getting ready for a pour—on my hands and knees cleaning up all the debris that the carpenters had dropped on that reinforcing iron, what we called the rat trap. I had a five-gallon bucket and was throwing in all these pieces of wood and crap. Right about then, somebody wasn't paying attention and the chute flooded with cement. It backed up over the top and spilled out from about 20 feet in the air right on top of me. They used that huge two-and-a-half rock in that cement, and everywhere the rocks hit me I got cut. I wasn't knocked out because I had a hard hat on, but everywhere I was cut, I got cement goo under my skin. My back was one big water blister.

The completed Marin pier stood 64 feet high and 44 feet above the water line, and measured 65 by 134 feet at the top. It was finished June 27, 1933, two months ahead of schedule. The engineering plans had worked perfectly and the job had been performed without a serious accident. But a celebration would have seemed premature. Still ahead was the south pier. The independent study conducted in 1926 by the Joint Council of Engineers had held that the foundation rock on the San Francisco side of the Gate was too unstable to support the weight of a 128,000-ton concrete pier and a 746-foot tower of steel weighing more than 44 million pounds.

Before the bedrock could be penetrated, the channel had to be conquered. Even a 128,000-ton concrete pier needed protection against the unrelenting tides of the Golden Gate. A protective ring was designed for the pier—a concrete fender that would act as a seawall, taking the brunt of the channel's force while allow ing the pier to be erected in calmer waters inside the fender. This fender, 155 by 300 feet and shaped like an overinflated football, would remain the pier's permanent bodyguard against storms and collisions.

Since the fender would be constructed 1125 feet from the San Francisco shore, an access trestle, or working wharf, had to be built into the bay. (The treacherous current precluded the use of traditional construction barges.) Before they could sink the supporting piles for the wharf, workers had to dynamite the bedrock at the bottom of the channel. By August of 1933 the wharf was nearly completed, and a guide frame for construction of the fender was in place. That year, the summer fog brought with it the bridge's first disaster.

During the night of August 14 a freighter trying to grope its way through the fog plowed into the middle of the trestle and ripped it in half. Even as the trestle was being rebuilt, a storm struck on October 31, then another on December 14. Howling winds turned the full fury of the channel against the trestle and the uncompleted fender. Bridge engineers totaled up the loss: $100,000 and five months work.

As a result of these losses, it was decided that the fender should be anchored a full 100 feet below the water, which required blasting 18 feet into the rock at the channel's bottom. In the spring of 1934, work began on the foundation for the fender. Since the blasting took place at depths of 100 feet, a new technique was devised to place the charges. A derrick barge was floated over the pier site and anchored securely—as securely as anything could be—inside the channel. A 14-inch-diameter pipe was lowered until it rested on the rock bottom, and a 5,000-pound, solid-steel shaft with a drilling point was repeatedly dropped through the pipe until successive impacts had gouged a hole two feet deep. Small dynamite bombs were then dropped down the pipe. Exploding upon impact, they blasted holes 18

(Right) A fleet of cement mixers provided a constant stream of concrete for the Marin pier. A total of 45,000 tons were chuted down an "elephant trunk" to workers in high rubber boots.

(Left) A fender was built 1125 feet from the San Francisco shore and anchored 100 feet below the waters of the Golden Gate. The fender was de-watered, forms were built and concrete poured until the north pier reached its final height 44 feet above the water.

(Below) The two anchorages weigh 50,000 tons each. The main cables consist of 61 bundles, or "strands" of steel bridge wire. One strand is fastened to each of the 61 eyebar chains embedded in the anchor block.

feet into the rock. Heavier assaults were unleashed by larger bombs, eight inches in diameter, 20 feet long and charged with 200 pounds of dynamite. The larger bombs blasted out great chunks of the channel floor to substantially widen the holes made by the smaller bombs. Then the blasted rock had to be brought up by clamshell dredger buckets. The excavated rock was so hard that the steel jaws of the dredger buckets had to be frequently replaced.

This underwater bombing and excavation was carried out from the deck of a barge under constant assault by high winds and waves, pitching the vessel up and down as much as eight feet. Even veteran barge workers became seasick.

Deep-sea divers were employed to set charges and, later, to help guide the forms into position for construction of the concrete fender ring. They had to work in virtual darkness at 100-foot depths, timing their dives to avoid the channel during its most turbulent moods. They were in constant danger of having their air hose and communications lines torn apart by the sudden riptides. The same violent bay currents could hurl a diver against the rocks or into the steel guide frames.

Diver Bob Patching remembered the perilous nature of the job:

It was treacherous as hell. You could only work at slack water, the period of high or low tide when there is no visible flow of water. . . . Even when it was slack, it could be tough. While it was slowing down or speeding up, it could still move fast enough to come awhipping in and knock you for a loop. I think we only had an hour and 15 minutes we could work down there at any one time. And then we would have to come up so fast we couldn't recompress in the water. The safest way to recompress is to do it naturally, on the way up. But what're you gonna do if the current is coming in so fast it'd knock you right out of the water?

They brought us up after a dive, and there wouldn't even be enough time for us to take off our suits . . . me and all my riggings hopped on back of a pickup so's it could speed us to the recompression chamber down there at the end of the dock. A few times I'd feel the bends coming on just before I made that chamber. But you get educated pretty fast on that subject. You know how long you've been down there. Sometimes you just have to make a run for it or you're a dead pigeon. That's one thing you don't crowd. That gets you too close to the old man.

Even as men risked their lives 100 feet below the Golden Gate, there was time for the kind of frolic which San Francisco loved. The legend that there were mermaid caverns beneath the waters off Fort Point had been cherished for years. Leaving nothing to chance, three men—eager volunteers—were chosen to dive off Fort Point and conduct an official investigation. Alas, they found no caverns and saw no mermaids. Fellow workers cheered their bravery nonetheless.

Fifteen days before Christmas, 1934, the bedrock foundation for the critical south pier was ready for inspection. Russell Cone, chief inspecting engineer, and Jack Graham, superintendent of the Pacific Bridge Company, were to make the inspection. They would descend to the bottom of the channel by climbing down a ladder inside an inspection well four feet in diameter and 107 feet deep. Before their descent they were bolted inside an air lock at the top of the long tube and the air pressure was increased to prepare them for the pressure at the bottom. A telephone connection linked the bottom of the tube with the surface.

Cone and Graham moved carefully down the well-lit ladder until they reached the bottom, where they stepped into an inspection bell, a dome-shaped steel chamber 15 feet in diameter and lit up like an ice cream parlor. Cone spoke first:

"Man oh man, look what great shape that rock is in!"

The engineers chipped at the hard rock and examined it with

PACIFIC BRIDGE CO

McClintic-Marshall

flashlights. They could not restrain their excitement.

"Why, this rock is great," shouted Cone. "Hard and firm! And look how clean it is—no loose rock to be removed."

Cone and Graham were all smiles as they climbed back up the long ladder and emerged into the sunlight. They pronounced the bedrock to be as solid as Gibraltar. The eight inpection walls and chambers could now be filled with concrete, and construction could begin. Joseph Strauss announced that completion of the south pier would be a New Year's Day present for bridge believers. It was actually the third day of 1935 before the San Francisco pier was topped off at 44 feet above the water, but it was declared complete on New Year's Eve and headlines extolled it as an "engineering miracle."

THE TOWERS

Now the San Francisco tower could begin to rise out of the waters of the Golden Gate. The Marin tower—completed on time, in late spring of 1934—was an inspiring sight. The tallest bridge ever constructed, it stood on a cliff at the perimeter of the Gate, while the San Francisco tower would stand *inside* the Gate, 1125 feet from the shore. North and south, it was the sight of the slender steel towers climbing into the sky that began to stir public excitement. The towers were the first visible evidence of the beauty to come. "A rising monument to God's will and man's ability and vision," read a typical headline, as photographers began competing for artistic honors with daily portraits of a masterpiece in the making. The Marin tower reached its full height just in time to greet an annual summer visitor drifting lyrically out of the Pacific. The new sentinels of the Golden Gate received the Pacific fog, and no engineer could have drawn on a blueprint the magic they would make together.

Their remarkably graceful stature belied the enormous strength of the towers, each constructed using some 21,500 tons of steel. The sections of steel were fitted together and fastened with rivets heated white-hot and shot through pneumatic tubes to the riveting crews above. Riveters worked inside cramped 3½-foot-square cells that honeycombed the interiors of the towers. Ventilation was poor in the cells, and the only light was supplied by small, unreliable lamps attached to the workers' hard hats. Every rivet was scrutinized by an inspector and if it was the slightest bit loose it was removed and replaced.

Another hazard faced by riveters was the danger of lead poisoning, a result of working in close proximity to the tower metal that had been heavily coated with red lead paint. Riveter Whitey Pennala remembered it as a near epidemic:

There were 60 guys in the hospital at one time, and nobody knew what was causing it. The doctors were diagnosing it as appendicitis—60 men all coming down with appendicitis at the same time. Finally it came out that it was lead poisoning. It was awful, that stuff. Guys were losing their hair, their teeth; they were breathing shallow. There was guys who never went back to structural work after getting a dose of lead.

Precautions were taken to prevent further cases of lead poisoning. Ventilation in the tower cells was improved; riveters were provided with filtration masks (some wore them, others refused); the lead paint was removed from riveting holes before rivets were driven; steel for the San Francisco tower was treated with iron oxide instead of red lead.

On May 4, 1934, an American flag was raised above the completed Marin tower. Before long, a ride to the top in the construction elevator became the biggest thrill in the West. When the Bay Counties Peace Officers Association met in Marin, the Sausalito police chief played host, taking a dozen fellow lawmen to the top of the Marin tower for an unequalled view of their territories. The officers took a priest with them and later reported

(Above) Ironworkers are a fearless breed, accepting the obvious risks and dangers as simply part of their job. They are nomadic, highly paid professionals who will travel anywhere to work on the high steel.

(Right) The towers were not built of solid steel but consisted of clusters of hollow cells. The base of each leg began with a honeycomb of 97 cells, gradually slimming to 21 cells at the top.

(Below) Prefabricated cells were lifted from barges and lowered into place on the towers. High up, a "connector" gave handsigns to the signalman below, who in turn sent light signals to the crane operator.

(Above) The roadway was constructed from the towers in both directions, carefully coordinated so the weight on the main suspension cables was evenly distributed.

(Left) Hot rivets touching red-lead-treated steel produced poisonous fumes. Filtration masks were required equipment for men working inside cramped, unventilated tower cells.

(Right) Bridge architect Irving Morrow added fluted corner brackets to tower portals. Their delicately drawn lines unmarred by rivets, the prefabricated brackets were barged from Alameda and installed in single complete units.

they were glad they had. The elevator became stuck on the way down and the party was suspended in midair for several moments.

The Marin tower was raised in less than ten months. Benefiting from experience, ironworkers erected the San Francisco tower in an astonishing 101 working days. But the task was not accomplished without interruption. Three weeks before the traditional "topping-off" ceremony, the partially completed bridge encountered its first earthquake.

Frenchy Gales was on the San Francisco tower when the temblor struck:

I guess there were 12 or 14 others up there too. I was walking across—I didn't hear nothing, and all of a sudden I felt like I was tipping off to one side. I sat down and shook my head. I thought I was dizzy or something. This friend of mine, he came along and asked what's the matter. I said, "Jesus, I feel like I'm swaying." He said, "It ain't you, the goddamn tower is swaying." I got up and sure enough, this whole structure was swinging like a hammock.

Guys were going crazy. Some were climbing down through the cells. One guy grabbed some old gloves, put them on, and slid down a derrick cable—took a chance on killing himself—those derrick lines were all greased up, you know. Most of us stayed up there while she was going way over, first to one side, then the other. The elevator that ran up and down outside the tower shaft was about halfway up and was swaying away from the tower, then coming back and banging against it. The poor guy inside was throwing up all over himself. Guys on top were throwing up too. After it stopped swaying we all went down as fast as we could, and got first aid for a sick stomach—I think it was a shot of whiskey.

Later, Clifford Paine reassured the public that the tower was not dangerously vulnerable to earthquakes. He explained that the San Francisco tower had swayed because the cables had not yet been attached. "As a result the tower acted like a tuning fork; it vibrated," Paine said. "But now, cables weighing millions of pounds are up, and it could never happen again." He was only partially correct.

The new sentinels of the Golden Gate were now in position. They became instant landmarks, immediate symbols of San Francisco. "Span Towers Seen in Napa" headlines exclaimed on a fogless day in the fall of 1935. Bridge District officials proudly reported that the towers could be seen on a clear day from the top of Mount St. Helena, 60 miles to the north.

Much of the credit for the clean beauty of the Golden Gate Bridge must be shared by Clifford Paine and staff architect Irving Morrow. If Joseph Strauss is to be memorialized as the man who built the bridge that couldn't be built, Paine and Morrow deserve to be remembered as the men who shaped it into the most beautiful bridge in the world. In direct contradiction of accepted design standards of the day, the towers gradually narrowed in width as they climbed upward. Said Paine, "Where great heights are involved, the common architectural practice of enlarging the scale of details as their distance from the ground increases tends to emphasize detail at the expense of the whole mass." The slenderizing effect was enhanced by the absence of the steel X-bracing in the open portals of the towers above the roadbed. There would be horizontal bracing only, leaving four open spaces in each tower and creating a visual sense of light and air which perfectly complemented the openness of the Golden Gate itself.

Strauss, whose final approval was needed on all design changes, accepted the new concepts advanced by Paine and Morrow, and praised the end effect as "a majestic doorway."

The towers' clean lines were given just the right flourish, just enough touch of art deco, to create an artistic work without succumbing to excess. Fluted brackets were fashioned for the

Footbridges 15 feet wide were suspended approximately three feet below the hypothetical line of the main cables. Crews moved along these redwood catwalks during cable spinning, sculpting, banding and wrapping.

upper corners inside the towers' open spaces, a notched design was applied to the horizontal bracing, and ascending vertical facets brought a sculptured effect to the outer legs of the towers. The bridge would stand as a singularly magnificent piece of work and still remain in harmony with its awesome framing of sea and land.

Thanks in great part to Paine and Morrow, Strauss would later be able to write, "It is a bridge built without gingerbread for decorative purposes. And the step-back towers are the simplest construction we could employ. Yet they have a grace and dignity that will endure, because of this simplicity, regardless of the way our ideas of architectural design may change from generation to generation."

THE CABLES

On August 2, 1935, the Golden Gate was closed to all shipping for the first time in history. Two Coast Guard cutters stood by to block traffic while a barge towed the first pilot cables across the Golden Gate. The pilot cables would support the midair catwalks from which the workers would spin the main suspension cables. The barge fought high winds and fog, taking nearly an hour to make the crossing. It took another four hours to raise the pilot cable. Once in place, it arched across the water as the first physical link between Marin and San Francisco.

The catwalks, narrow footbridges across the Gate, were a major construction feat in themselves. They served as scaffolding would on a building, except that this scaffolding had to be built across an open channel a mile wide and swept by wind and fog. Two walkways 15 to 18 feet wide were constucted with redwood-plank flooring and supported by thick steel-wire strands. Each catwalk was suspended about three feet below the planned path of the main suspension cables.

The catwalks were finished in three months and later strung with red, white and green lights that grateful Bay Area residents accepted as the Bridge District's thoughtful observance, a few weeks early, of the Christmas season. (The fact that the lights were installed as warning signals to ships and aircraft was largely ignored.) At the same time the San Francisco Bay Exposition announced a new sightseeing cruise that would take customers on a closeup tour of the first physical spanning of the Gate. The $1.25 cruise ticket included a luncheon buffet.

With the towers up and the connecting catwalks completed, workers began laboring under circumstances resembling those endured by high-wire acrobats. Spectators on the shores, or on the deck of the new cruise boat, marveled at the sight of men working high above the channel. Climbing about on the steel towers and swinging catwalks, they were frequently enveloped by fog and buffeted by winds. The towers swayed enough at the top to make workers motion sick, and it was bitterly cold. Ironworkers are accustomed to doing their jobs at heights that would panic most people. A city official visiting the job site looked up at the men scampering along the catwalk and remarked, "These fellows must have nerves of steel." "Hell no," the foreman replied. "If they had any nerves at all they wouldn't be steelworkers." But even the gritty ironworkers had trouble adjusting to the Golden Gate weather.

Ironworker McClain had helped to build the Empire State Building and the San Francisco-Oakland Bay Bridge before he went to work on the Golden Gate Bridge in the summer of 1936. Nothing had prepared him for the weather:

I think I was colder on the Golden Gate Bridge than I'd ever been in my life. Weather on the Bay Bridge wasn't bad at all. But on the Golden Gate Bridge, it was murder. You couldn't dress against it. Besides our regular clothes, we had on overalls and a jumper. When I first went out there I was freezing to death and one of the men brought me out an overcoat. And I put that on. Over all the clothes I already had on. And I still was cold. It was

a different kind of cold than we men who came out of the East were used to. It's raw cold right off the sea. It got to us. I was miserable cold.

Now on the Sausalito end, that was another story. Wind didn't blow across the quarter-span back there like it did out in the center. We had days that were quite balmy.

In the fall of 1935 the complex job of cable-spinning began. John A. Roebling & Sons of New Jersey had been awarded the cable-spinning contract for $5,855,000. The name John Roebling had been synonymous with cable-spinning since he built the pioneer suspension bridges in the second half of the nineteenth century. By 1935 the Roebling company had the intricate cable-spinning process down to perfection.

Unreeling machines operated from each end of the bridge and enormous spinning saddles were mounted atop the towers. A spinning wheel driven by electric motors unraveled wire from the unreeling machines, carried it across the catwalks and through the spinning saddles atop the towers. The cable wire was compressed into strands. It took 61 strands to make up one main cable; an average of 452 wires went into each cable. Although the many complications of the cable-spinning process are difficult to explain in layman's terms, the figures speak for themselves: 80,000 miles of wire was spun between the anchorages, a total of 27,572 individual wires, compacted into cables 36⅜ inches in diameter, each cable 7,650 feet long. The twin cables, suspenders and accessories weighed 24,500 tons.

Roebling crews completed the world's biggest cable-spinning job in six months and nine days, setting the final strand in place on May 20, 1936. At one point 420 men were spinning 735 feet of cable per minute. On good days 1,000 miles of wire were unreeled into place in eight hours.

THE ROADWAY

The Golden Gate had now been spanned by cables. All that remained was to cross it with a steel-supported concrete highway. This last piece of work would be the most dangerous. The "dance of danger" (as newspapers liked to describe the ironworkers' daily labor atop the towers and on the swaying catwalks) would now be performed on a bed of steel. The steel inched out from each shore and in both directions from the San Francisco tower, 220 feet above the merciless waters of the Gate.

Workers had reason for apprehension by the summer of 1936. Months earlier, the papers had begun to carry reports of how the Golden Gate Bridge was cheating "the law of the concrete and steel—one life for each million dollars." This has been the accepted fatality rate among bridge workers for years: for every million dollars spent on construction, one life lost. Ironworkers expected it. In fact, 22 men had already been killed during three years of construction on the San Francisco-Oakland Bay Bridge, where the weather was comparatively mild. The Golden Gate Bridge had been under construction for 3½ years, more than $20 million had been expended, and not one life had been lost. There had been injuries, some serious, but none fatal. The deadly equation hung over the project, raising the same question in everybody's mind: How long could such luck hold?

Of course it was more than luck that was responsible for the unprecedented safety record. Bridge engineers, from Joseph Strauss down, took extra measures to promote safety, to try and cheat the death-rate law. Because crew bosses were so strict about making workers wear hard hats, many assumed it had never been done before. It had, but at the time hard hats were such a novelty that the newspapers ran pictures of men going to work in what looked like "war helmets."

Everything possible was done to improve safety. In the beginning, many workers had to be treated for eye irritation caused

Sculpting machines applied 4,500 pounds of pressure per square inch and compacted the 61 cable strands into perfectly round main cables measuring 36¹⁄₁₆ inches in diameter. Sculpted cables then received cable bands around which hung the suspender ropes, to which in turn the roadway was fastened.

During construction of the roadway a safety net saved lives, tools and time. Workers, less fearful that an accidental slip would be fatal, moved with confidence and speed along the high steel. The trapeze-like net consisted of six-inch squares of ⅜-inch manila hemp and extended beyond the work area.

by rapidly changing light levels on the catwalks and towers. Workers would be groping in fog one moment, then blinded by dazzling sunlight the next. After intensive testing in the various light conditions, workers were fitted with the same filter glasses worn by Navy gunners and fliers; it was another first for the bridge.

A well-equipped field hospital was located at the job site. A male nurse was on duty inside the little shanty at all times. By the end of 1935 some 12,000 cases had been treated there. "And there has not been one instance of infection," reported the press. "That is declared to be another world record."

Workers were reminded at every turn that attention to safety was an important job requirement. One placard noted that 266 persons were killed every day in the U.S. by accidents. Another quoted Irvin S. Cobb: "I'd rather be late for dinner tonight than be on time for breakfast in the next world in the morning."

Safety regulations were rigorously enforced. A man caught showing off on the towers, on the catwalks or on the roadbed could be fired on the spot. Safety belts, known to the ironworkers as tie-off lines, were required equipment, along with the hard hats and the filter glasses. Ironworker McClain recalled the high priority placed on safety:

We had very strict safety rules. You didn't do any showoff business. That was a way to get yourself kicked off the job, if you started grandstanding. The companies were pretty safety conscious. You see, we introduced out here quite a few safety measures that had never been used before. We didn't have a net on the Bay Bridge, but we had a net on the Golden Gate. And tie-off lines. On our belts we had a piece of line about eight to ten feet long. If you got to one spot where you were going to be working for a few minutes and had a chance, you tied off your tie-off line. It saved many lives.

The most effective safety device was new to bridge building but as old as the circus: a trapeze net. The safety net cost $82,000, plus $48,000 to install. It draped 60 feet below the roadbed and extended ten feet out on either side. The net worked so well that newspapers began running box scores: "Score on the Gate Bridge Safety Net to Date: 8 Lives Saved!" They also reported the formation of an exclusive club by men who had fallen into the net—the Halfway Club. Manners of the day prevented the papers from printing the name that the men had actually given their exclusive fraternity: the Halfway to Hell Club. At times it seemed that a plunge into the safety net was nothing more than a sporting diversion to the high-wire work that ironworkers took as routine. "It wasn't half as much of a thrill as the time I fell seven stories off a construction job in New York," said riveter J.E. Roberts, a newly installed member of the Halfway to Hell Club.

Reporter Kenneth McArdle was understandably agog at the nonchalant attitude:

Hardly had they [two men who had earlier fallen into the net] got back to their work when a body hurtled past them and the net sagged once more. This time it was . . . a painter whose foot had slipped, plunging him backward from a scaffold. He fell 45 feet, bounced, and climbed out with no further injury than a twisted ankle. Yesterday another painter followed . . . [he] toppled off a steel cord at the middle span, landed in the mesh, crawled out, and went back to work . . . No. 8 on the roster of men for whom a flimsy-looking web mesh, spun beneath the giant span, has measured the difference between life and death.

Actually, although the net unquestionably saved many lives, it also added to the number of men falling because they worked with freer abandon, knowing the net was below them. "The unofficial record of men who fell into the net was 18," said Harold McClain. "Many of them would not have fallen had the net not been there. . . ."

A work force that peaked

Even before the roadway was paved, the endless task of painting the bridge had begun. Squeezed into bosun's chairs, painters slowly hand-brushed the suspender cables with international orange. Today rusted and chipped paint is quickly sandblasted away and fresh coats of primer and vinyl paint are applied with spray guns.

at 1,300 men had labored under extreme danger for three years, eight months and 16 days before the Golden Gate Bridge claimed its first life. On October 21, 1936, a derrick lifting steel roadway beams toppled and crushed Kermit Moore, a 23-year-old San Franciscan employed by Bethlehem Steel. Another worker was knocked off the steel flooring and fell 35 feet into the net. McClain was there when the first death occurred:

Kermit had just hooked onto the cord section and for the moment his buddy was off somewhere else doing something, so Harold Horton had gone over and was helping him force the choker to choke dead center. Harold was an old-timer at the business and when he saw the derrick coming toward him he dropped below the cord and partly under the car. Kermit was on his first raising gang job. Horton reported that he was looking directly at Kermit as the derrick came over, but Kermit froze on the spot and could not move. One leg of the traveler landed squarely on the part of his body that was leaning over the cord, and that part of Kermit was destroyed as completely as an insect killed with a club. Horton and I volunteered to find as much of Kermit as we could. . . .

The derrick accident was a grim reminder of the dangers faced daily by bridge workers. Still, one death in nearly four years of work was an impressive safety record.

Joseph Strauss personally operated the derrick that lifted the final piece of steel flooring into place on November 18, 1936. The end of the great task was so near that even the wearied bridge workers were unable to conceal their excitement. The steel arms reached out from the opposite shores until they connected in the center.

The last piece in the picture, the suspended roadway, was being fitted into place as smoothly as leaves into a long dining table. Suspender ropes were dropped down from the main cables to await the load of the roadway. The roadway was manufactured by sections in Eastern steel mills, shipped to Alameda, then carried by barge across the bay. Derricks mounted at the floor level of each tower raised the steel to the roadway level. Traveling derricks then carried the road sections into position, starting at the towers and progressing across the channel as the steel was riveted into place. After a series of 90-foot cross beams were topped by longitudinal stringers running parallel with the axis of the bridge, and everything was riveted into position, the concrete roadway was poured. Forms for the concrete were suspended from the stringers and the concrete poured in rectangular slabs. A total of 24,000 tons of steel and 25,000 cubic yards of concrete went into the suspended structure that would at long last open a road between San Francisco and the North Bay counties.

As the roadway neared completion, the city was in a joyous mood. Bridge believers—their numbers swelling in proportion to the visible progress of the bridge—made sport of those who had said it couldn't be done. "Reminder," was the glib headline on a short editorial in the *San Francisco News*: "Every structural part of both the Golden Gate and Bay Bridges that could possibly interfere with navigation has long been in place. And we ask our readers to note that once more the greater part of Uncle Sam's Navy has entered the gate and steamed to its anchorage without a single bump! P.S. And the south pier of the Golden Gate Bridge hasn't slipped into the Pacific yet!"

Euphoria was building by the day. It was abruptly interrupted on the morning on February 17, 1937. Harold McClain was replacing rivets on the Sausalito end and he's never forgotten the moment he knew something was wrong: "I felt the bridge tremble." Loud crashing noises were followed by the ominous sound of something being torn, like a giant bolt of cloth being ripped down the middle by unseen hands. But the thunderous sound was made by something all too visible. A special scaffold that moved on rollers had been suspended beneath the bridge roadway so that men working from the scaffold could remove the forms used to pour the concrete paving. The rolling scaffold was being used that morning for the first time. Inspectors, wary of the scaffold's stability, were in fact on their way to test it at that very moment. They were too late. The scaffold tore loose from the underbelly of the bridge, the rollers gave way and the scaffold crashed into the safety net, carrying 12 men with it. The scaffolding platform weighed ten tons. It hung there in the net for a few agonizing seconds before the net began tearing apart under the weight. The scaffold plunged another 250 feet into the channel at one of its deepest points, taking parts of the net and 12 men with it. They landed in a roaring tide that was surging through the Gate and out to sea.

Two men survived. One body was recovered. Boats searched in vain for the bodies of the other nine workers, following the tide out the Gate and two miles into the ocean. At least four of the men were seen struggling to stay afloat by clinging to scraps of wreckage from the scaffolding. It was an impossible battle.

No one was sure how many men had been working on the scaffold, or how many had been swept out to sea. As each workman returned to the field office and deposited his time card, his name was removed from the list of possible victims. After all the

Originally scheduled for destruction, Fort Point was saved by Strauss who decided to build over and around it. The steel arch now frames the historic landmark which is visited by more than a million people annually.

workers had reported in, the slots for ten time cards remained empty.

One of the bridge divers, Bob Patching, arrived at the Fort Point construction office with his diving gear. "My pal, Shorty Bass, is somewhere at the bottom of the Gate. I want to go down and bring him up," he said. Crew bosses sympathized, but told him it was hopeless. If Shorty had been trapped in the net, and not carried out with the tide, he would be at the bottom of 350 feet of water, and nobody could dive that far down. "Shorty's down there," Patching insisted. "I've never gone that deep before, but I'll go. Maybe if I can't find Shorty, there'll be others down there. . . ."

Bridge worker Peter Anderson had watched helplessly as his brother, Chris Anderson, tried to climb out of the net. They found his body a few days later, still wrapped in sections of the torn net.

The newspaper accounts of the accident would have seemed turgid if fictionalized in a dime novel. But there was no way reporters could exaggerate the scene of prolonged horror, which was most chillingly described by the workers themselves.

The cries of the men seemed to split the air. The weight of the crane and falling net shook the bridge like an earthquake. There was a 10- or 15-mile west wind, and the net waved like a huge flag as it carried the men down, away from us. I shall never forget the contorted white faces fading away. (Painter Roy Trimble)

I felt the tower tremble as though there were an earthquake. I could see the net falling, accompanied by a sort of subdued chatter. I could hear faint, babylike cries. When the net hit the water the men seemed like little blots of ink on the surface. The net looked like a rapidly sinking raft, with the tiny men entangled, fighting to get free. Then some of the inky blots disappeared. Some drifted out the Gate and out of sight. (Ironworker Tex Leaster)

The two men who survived the fall were E. C. "Slim" Lambert, a 26-year-old laborer, and Oscar Osberg, a 51-year-old carpenter. They were saved by a fishing boat that was 100 yards away when they hit the water. Osberg suffered a fractured hip, broken leg and serious internal injuries. Lambert was virtually unhurt.

A year-long investigation followed. Workers had maintained from the beginning that the scaffold platform wasn't properly bolted into place. Joseph Strauss and Clifford Paine blamed the contractor, Pacific Bridge. The contractor said the engineers were trying to cover their own mistakes in failing to properly supervise the job. The Industrial Accident Commission investi-

The slender towers are made more elegant by the absence of x-bracing above the roadway. The color, international orange, was not an immediate favorite, but it is now hard to imagine the Golden Gate Bridge painted any other hue.

gated and returned a verdict against undersized safety bolts, but declined to place blame. Bridge workers remained convinced that the accident was caused when safety was sacrificed for expediency in a rush to complete the job.

The tragedy may have marred the completion of a monument to industrial safety, but even so, the Golden Gate Bridge had set new standards of safety. The law of the concrete and steel, one life for every million dollars, had been broken: the bridge had cost $35 million and 11 lives.

The Bay Bridge opened first, on November 12, 1936. But it was already evident that it would forever be San Francisco's number two bridge. Stunning engineering achievement that it was, and structurally handsome in its own right, the Bay Bridge simply did not compare to the splendor of the Golden Gate. It seemed more a celebration of hard-fought progress, a vital connection beween San Francisco and the growing city of Oakland and the surrounding East Bay counties. The Golden Gate Bridge, though similarly motivated by man's desire to move freely between opposite shores, became a celebration of beauty in harmony with advancement of commerce.

Bay Area motorists were given a preview of what to expect by intrepid *San Francisco News* reporter Anna Sommer, who scooped her rivals by hitching a ride across the Golden Gate Bridge with Strauss and bridge district president William Filmer a month before the official opening. Like the millions who would follow her, Sommer immediately felt the excitement of crossing San Francisco's other bridge recede in the face of this exhilarating experience:

If you think you've seen something after zooming across the San Francisco–Oakland Bay Bridge—just wait until you drive across the Golden Gate Bridge. It's like a sky-ride—like viewing the world from a crow's nest on a ship—like looking down on a Norwegian fjord—like being on top of the Matterhorn—oh, well, we give up. There aren't similes enough to describe it.

It wasn't like driving over the Bay Bridge, where you are hardly aware when the long approach gives way to the suspension span—and even then it's more like rolling along the Bayshore Highway, because the solid concrete balustrades cut off the immediate water view. A couple of hundred feet and you're right over the churning tides of the Golden Gate. You know it because the balustrades are like a steel picket fence—with the pickets wide enough apart to give you the dizziest sensation you've ever known outside an airplane. Alcoves are set at intervals in the balustrades. Mr. Strauss said he had planned to place benches there for pedestrians, but changed his mind because it would make it too easy for suicides.

If you're one of those who fumed because the Golden Gate Bridge is painted "international orange" instead of aluminum like 1the Bay Bridge, you're in for a breathtaking surprise. The "orange" is a subdued terra cotta that resembles the rich earth tones of the Grand Canyon, contrasted against the green Marin hills, the blue sky and bay.

THE COLOR

Indeed, there had been considerable "fuming" about the color of the bridge. International orange was not an early favorite and its ultimate selection aroused more than a little protest. "A lot of ladies are raising a fuss against the international orange color on the bridge and demanding aluminum," was a report received by bridge district directors in the summer of 1936. Petitions bearing hundreds of signatures protesting the color were delivered to directors.

Few people know that the Golden Gate Bridge might have ended up looking like a rainbow popsicle. One of architect Irving Morrow's early color schemes called for four shades of red, and possibly yellow for the cables. While sensitive to the choice of color, Strauss and his aides were primarily concerned with finding a paint that would withstand exposure to severe weather and resist corrosion by the salt air that was constantly attacking the steel. After a year of testing hundreds of paint samples, the red-and-yellow color scheme was rejected, and the choices narrowed to steel gray, carbon black and orange-vermillion. Morrow, having earlier espoused the use of red, leaned more to the warmer orange color, correctly judging the dark or gray colors as too somber for such an illustrious setting.

Exactly how international orange was finally selected remains clouded. Some say it won out simply because it tested best in the resistance to rust and corrosion, and because it was the closest choice to red lead paint possible. But by the time the bridge was completed, international orange was generally praised as the right choice. One of San Francisco's favorite artists, Beniamino Bufano, was an early advocate of the orange (or terra cotta) shade. "Let me hope that the color will remain the red terra cotta because it adds to the structural grace and because it adds to the great beauty and color symphony of the hills," Bufano said in a letter to bridge officials. It would soon be impossible to imagine this symphony of hills, sky, ocean and fog being in tune with a steel-gray Golden Gate Bridge.

The traditional gold rivet marking the completion of the Golden Gate Bridge was to be driven on April 28, 1937. Politicians, bridge officials and hundreds of others converged at mid-span. The ceremony got off to a shaky start as two companies from the Army's Sixth Coast Artillery marched onto the span. A frantic bridge official, perhaps remembering a bridge disaster some years earlier that was blamed on vibrations caused by troops marching in step, ordered them to stop. They did and the ceremony continued.

The rivet was allegedly made of Sierra gold, but seasoned ironworkers and riveters observed with amusement that the "gold" rivet was liberally laced with brass.

The honor of driving the final rivet was bestowed upon the man who had driven the bridge's first rivet. He was Iron Horse Stanley, tenth member of the Halfway to Hell Club. A cheer went up as Stanley's gun struck the rivet, but the next day's *San Francisco Chronicle* reported: "That gun was made to hammer down steel and not soft gold. Fine particles of gold showered in the faces of the spectators. Warden Court Smith of San Quentin Prison even getting a bit half the size of a fingernail for a souvenir." The soft gold rivet fell apart and the hole was plugged with an ordinary steel rivet. A gold rivet was later installed elsewhere.

It was done. The Golden Gate had finally been conquered. But not without a fight.

The bridge had had its detractors. A national Depression had nearly killed it. Daring men had fought with dynamite bombs and hot steel above and below the Golden Gate to see it through. There had been casualties, and 11 deaths—but a relative handful of men had fought the odds and won. And they had brought forth the most beautiful bridge in the world.

Wrote Joseph Strauss:

The Golden Gate Bridge,
the bridge which could not and should not be built, which the War Department would not permit, which the rocky foundation of the pier base would not support, which would have no traffic to justify it, which would ruin the beauty of the Golden Gate, which could not be completed within my cost estimate of $27,165,000, stands before you in all its majestic splendor, in complete refutation of every attack made upon it.

Verbose with a pen, the hard-boiled engineer with the soul of a poet and the fight of a field general would later respond more concisely. As everyone around him went mad with excitement, he acknowledged with uncharacteristic modesty the culmination of his life's work, and the realization of a great dream: "This bridge needs neither praise nor eulogy, nor encomium. It speaks for itself."

3

THE ENDLESS CELEBRATION

(Left) Opening Day, May 27, 1937, was Pedestrian Day. During the first hour 19,300 people passed through the San Francisco toll gates; 5,000 more came across from Marin. By the end of day, an estimated 200,000 citizens had made the crossing on foot.

(Right) Nearly 18,000 people were poised at the barriers for the early morning opening. At 6am the foghorns bellowed, the toll gates lifted, and one of the wildest eruptions of human exuberance in the history of the West was unleashed.

The Golden Gate Bridge was built for automobiles, but its seductive powers draw people to it for reasons more personal, more fanciful, and often more bizarre, than the mere expediency of using it to cross the bay. It is a bridge that was bought and paid for by the people, by choice, not built for them by the government, and the people claimed their purchase with instant fervor. If history needed a precise moment when it became the people's bridge, it would logically settle on 6am, May 27, 1937, when the bridge was opened to pedestrians. But perhaps the date should be fixed 11 hours earlier, when 15-year-old Boy Scout Walter Kroneberger unrolled his sleeping bag in front of the No. 3 toll booth and prepared to spend the night as the first person in line to walk out onto this long-awaited bridge. This opening encounter between the people and their bridge was a prophetic one. The toll captain invited the boy inside one of the toll booths and he slept in his sleeping bag on the floor, warmed by an electric heater. It was the first example of what would become a strong personal relationship between the people who operate the bridge and the people who use it.

San Francisco, a party-loving city from the day Sam Brannan first rode into town with a fistful of gold, decided that the opening of the Golden Gate Bridge was an event too monumental to be confined to just one party. They set aside a week to celebrate. The Golden Gate Bridge Fiesta was one long eruption of jubilation shared by a citizenry bursting with pride. A Fiesta Queen was crowned and escorted to her throne by Hollywood star Robert Taylor. Songs were composed and poems written in homage to "this jeweled necklace now draped across the Golden Gate." Merchants dressed in Gold Rush costumes. A dramatic pageant, "Span of Gold," starring a cast of 3,000, played nightly at Crissy Field "in the world's largest outdoor theater."

A torchlight parade drew more than a hundred floats and thousands of marchers. Fireworks filled the sky above the bay, Al Jolson sang and people danced in the streets. There were bowling tournaments, log-sawing contests, boat races, marathon relays, a wild west show and, fittingly, a hard-rock drilling championship. And of course there were innumerable chain-cutting, bridge-dedicating and speech-making ceremonies. Even the approach to the bridge deserved a dedication of its own: the Redwood Empire Association officially opened the Waldo approach from the Marin side in a "Hands Across the Golden Gate" ceremony. The U.S. Navy timed the return of a part of its fleet from maneuvers in the Pacific to coincide with the opening festivities. A formation of 500 Navy planes roared above the new guardians of the Golden Gate, heralding the fleet's arrival.

All of these celebrations combined did not have the impact of one 12-hour period between 6am and 6pm on the cool, clear day of May 27. Recognizing this as one of the most exciting days in San Francisco history, bridge officials wisely chose to follow an example set by the opening of the Brooklyn Bridge in 1883. For those first 12 hours, the Golden Gate Bridge would belong completely to the people. Automobiles would have to wait for the *second* day.

By 6am, the announced starting time, 18,000 people were massed on the two sides of the bridge, the vast majority at the San Francisco bridgehead. Foghorns bellowed, the toll gates opened and one of the wildest eruptions of human exuberance in the history of the West was unleashed. During the first hour, 19,300 people streamed through the San Francisco toll gates; 5,000 came across from Marin. People moved up and down the roadway sampling the views of the bay, the ocean, the San Francisco skyline, Alcatraz, Angel Island, the Marin hills, the old forts guarding each flank of the channel—it was a feast of views consumed for the first time from this magic platform 220 feet above the waters of the Golden Gate. Reverently, they touched the steel, fingered the rivets, put their hands on any part of the bridge they could

The Golden Gate Bridge Fiesta was a week-long celebration. A queen was crowned, merchants dressed in Gold Rush costumes, a pageant with a cast of 3,000 played nightly at Crissy Field, fireworks filled the sky. Al Jolson sang and people danced in the street. There was a torchlight parade, races, tournaments and contests. Songs were composed and poems written to "this jeweled necklace now draped across the Golden Gate."

reach, as if to prove to themselves that it was real. They spat over the side and shuddered at the sight. "Fishing boat!" someone would yell, and the crowd would scramble to the railing, waving and shouting to the fishermen passing under this manmade miracle.

The bridge had a hypnotic effect on people and inspired a stirring of the imagination that would only grow stronger through the years. Merely crossing the bridge, even on the historic first day, was not enough. There was a human comedy of Golden Gate Bridge firsts: first baby across in a stroller; first brothers to cross; first person to take a picnic across to the Marin hills; first person to walk across on her birthday; first roller-skaters across; first dog to walk across (a dachshund named Fritz, and history recorded that he made a nonstop crossing despite all the lamp poles); first priest to cross; first person to cross on stilts; first person to walk across with her tongue sticking out. The first person to fail in a bridge stunt was the man who quit after 100 feet of pushing a pill box across the bridge with his nose. A highway patrolman became the first person injured in an accident on the bridge when he crashed his motorcycle into a barrier. The first barefoot crossers (two young men whose shoes were thrown over the side by friends) also became the bridge's first hitchhikers when they thumbed a ride back on a passing ambulance. Children were lost, and assorted valuables and debris were dropped, or purposely thrown, into the safety net still draped under the roadway. (Officials declined the offer made by a professional high-diver from St. Louis, who wanted to commemorate the opening by diving off the bridge.)

The pedestrian traffic was so heavy that the turnstiles broke down. By 6pm, when the toll gates closed, an estimated 200,000 people had swarmed onto the bridge, streaming through the toll gates at the average of nearly 200 per minute. It was an auspicious debut for a bridge that would draw more world attention than any other manmade structure in America. It was the beginning of a celebration that has never ended.

Exhausted bridge officials and highway patrolmen, relieved that Pedestrian Day had passed without serious trouble, braced themselves for the bridge's next step into history—the arrival of the automobile. In contrast with the first crossing by pedestrians—which had been a display of high spirits and near-pandemonium—the opening day for the automobile began with a decorous procession led by San Francisco Mayor Angelo Rossi, Joseph Strauss, and a caravan of high-ranking bridge engineers, district officials, politicians and V.I.P.s from throughout California, Canada and Mexico.

Joseph Baermann Strauss was no ordinary civil engineer. He was a technician and politician, a dedicated visionary who, in the face of countless obstacles, pursued a dream for fifteen years. Sadly, less than one year after his Golden Gate Bridge was opened, Strauss died of a heart attack in Los Angeles.

At 9:30am the official convoy of black automobiles rolled slowly down the Waldo approach on the Marin side, drove through a 16-foot-long redwood log that had just been sawed in half for the occasion, stopped for the ceremonial torch-cutting of successive chains of copper, silver and gold, then moved in funereal columns out across the bridge. Greeting them on the San Francisco end was a row of Fiesta Queens in ruffled gowns and flowing Spanish veils. The motorcade stopped for the final bridge ceremony, and Joseph Strauss officially presented the bridge to district president William Filmer, his final act in a personal drama that had begun 20 years earlier. It was a brilliant climax to the career of a master bridge builder whose greatest accomplishment has never been surpassed.

Strauss moved to Los Angeles after the bridge opened and died ten months later at the age of 68.

The Golden Gate Bridge was declared officially open at noon May 28, 1937, when President Franklin D. Roosevelt pressed a telegraph key in the White House announcing the event to the world. San Francisco responded with the loudest burst of noise-making in its history. Foghorns blared, church bells rang, police and fire sirens screamed, ships in the bay blew their whistles, drivers throughout the city honked their horns, firecrackers exploded, and anyone who didn't have a noise-maker just whistled and shouted.

During its first official hour of business, the bridge handled 1,800 cars. Pedestrians still outnumbered the vehicles: 2,100 walked through the turnstiles. The procession continued through the night, and by midnight an estimated 25,000 cars and 19,350 pedestrians had paid their tolls to cross the bridge (5 cents for walkers, 50 cents for drivers).

The city's celebration continued for another five days. But on the bridge the party had ended and the serious business of operating the world's longest, tallest suspension bridge began. Skeptics were still waiting for the bridge to prove itself on two counts: (1) Would there be enough traffic to make the bridge financially self-supporting? (2) Would it stand up against earthquakes and storms? There was no way to anticipate when, or how severely, the bridge might be tested by earthquakes or Pacific storms. The traffic count, however, was monitored as closely as the daily attendance at a World's Fair.

During the first full fiscal year, from July 1, 1937, to June 30, 1938, a total of 3,326,521 vehicles crossed the Golden Gate Bridge, generating $1.6 million in tolls. The income would have been higher if not for some 100,000 "free-riders" who benefited from the War Department policy that guaranteed federal employees toll-free use of the bridge. The bridge district howled that it was losing hundreds of thousands of dollars, as government employees used "working" passes to take family and friends on joy rides across the bridge. It took an act of Congress in 1944 to end the practice, restricting the free passes to uniformed personnel.

The ferryboats were presumed obsolete once the bay had been spanned by the Golden Gate Bridge and the San Francisco–Oakland Bay Bridge. But the ferries did not surrender peacefully. After the Golden Gate Bridge opened, the Southern Pacific Ferry Company tried to forestall the inevitable by cutting fares. This ignited a price war between the ferries and the bridge, but the skirmish ended in a matter of days, and the bridge district, knowing that it held all the aces, raised auto tolls back to 50 cents. The Southern Pacific ferries that had carried people and their cars from Marin to San Francisco for so many years were gone by 1941.

In the early years the Golden Gate Bridge was scoffed at by fiscal conservatives as a bridge built for weekend pleasure driving. They portrayed the $37-million span as a recreational novelty, not the great conveyor of daily commerce that had been promised. The bridge district bought newspaper and magazine space to promote travel to the "Redwood Empire," the pop-

JIM BEAM

ular appellation given the North Bay counties. Before long bridge officials would look back nostalgically on the days when they had worried about the bridge attracting enough traffic.

WAR!

Retired bridge officer Ed Moore remembers sitting in the sergeant's office on the morning of December 7, 1941, when he heard the news. He called the captain to tell him that the Japanese had just bombed Pearl Harbor. "The captain laughed and thought it was a joke," Moore recalls. "Then I looked out the window and saw a highway patrolman running out on the bridge with a rifle, looking up in the air." Army and Navy officers could soon be seen standing along the west railing of the bridge, scanning the horizon.

All the old fears planted by the War Department came rushing back. If the Japanese were bombing Pearl Harbor, wasn't it possible that their next move would be to strike at West Coast naval defenses? San Francisco was a natural target, and suspended across the entrance to the harbor were millions of tons of steel and concrete, enough to block the channel and bottle up the fleet if the bridge were bombed. There was little solace in Joseph Strauss's old argument that if enemy bombers were ever able to destroy the

The San Francisco-Oakland Bay Bridge, seen above, was completed and opened six months before the Golden Gate Bridge. An efficient and honest structure in its own right, it has never generated the popular affection inspired by its sister span to the west.

bridge, they would also be able to just about wipe out San Francisco. National Guardsmen were sent to the bridge, where they remained on alert throughout the war.

The Golden Gate had always represented the open arms of welcome to San Francisco. Now it was the exit to uncertainty for millions of troops as they shipped out to fight the war in the Pacific. Troop ships steaming slowly underneath the bridge were often hidden by the blackness of night or moved in a shroud of fog. Bridge workers could hear soldiers shouting their goodbyes, goodbye to San Francisco, to the U.S.A., to a wife or mother, or just goodbye to the Golden Gate Bridge, which they expected to be there to greet them when, if, they returned.

And when they returned, the sight of the orange steel towers and arching cables became a symbol. Throughout America, soldiers returned to their families and friends repeating the same story: "I knew I'd made it, I knew I was back home, when I first saw the Golden Gate Bridge."

A BUMPER CROP

The war had effectively ended America's Depression. Post-war prosperity followed. Those who had moved to the Bay Area to work in the shipyards stayed. Men who had passed through on their way to the war in the Pacific returned. There were plenty of jobs, good salaries, and cheap housing in the sprawling suburbs. Everybody owned a car, some people owned two. The Golden Gate Bridge and Highway District had no further need to advertise for business.

During 1946, the first full year of peacetime following the war, total bridge traffic topped 6 million vehicles for the first time. By 1951 the count was over 10 million, and that figure had doubled by 1961, when 20,011,302 vehicles crossed the Golden Gate Bridge.

The bridge was rapidly becoming outdated as a traffic carrier. The North Bay counties, especially Marin, had grown faster than bridge proponents could have imagined. San Francisco remained the core of the job market, but escalating housing costs, high crime and deteriorating public schools sent families across the bay in search of better living conditions. As the populations of Marin and Sonoma Counties increased, the lines of cars crossing the bridge during commute hours grew ever longer. It was small comfort to drivers stuck twice a day in rush hour traffic that the jams occurred on the world's most beautiful bridge.

There was talk of building a second bridge between San Francisco and Marin, but San Franciscans rebelled at the idea. Spoiling the beauty of the internationally famed Golden Gate Bridge by building another span on the same site was unthinkable.

An alternative proposal—a second deck to accommodate high-speed commute trains—was less controversial, but still generated considerable opposition when it was first introduced in 1956. The mere suggestion of tinkering with the Golden Gate Bridge inevitably brings howls of protest from traditionalists, and the vision of a second deck met with predictable disfavor. Nevertheless, the ensuing debate lasted for 11 years. Official estimates for a second deck—$35 to $40 million—were higher than the original cost of building the bridge. The second-deck concept began to collapse under the weight of public opinion, and in 1967 bridge district directors conceded defeat.

Two years later, in an attempt to alleviate the worsening traffic problem, the state legislature passed an act that gave the bridge district new authority and new options. The renamed Golden Gate Bridge, Highway and *Transportation* District suddenly found itself in the bus and—height of irony—ferry business.

In fact, ferries had already made a comeback on San Francisco Bay when a privately owned company, Harbor Carriers, had begun operating a fleet of excursion boats in 1962. Although Harbor Carriers catered primarily to tourists, the company also inaugurated a commuter service between Tiburon and San Francisco. Response was tepid; only a few hundred commuters used the ferryboats on a daily basis.

The bridge district, however, was committed to making the ferries part of the Bay Area's commute network. In 1970 the distict's first vessel, purchased in San Diego, began to carry passengers between Sausalito and the San Francisco ferry terminal. Seven years later three more boats, built to order at a cost of $4.2 million each, were carrying passengers between San Francisco and the town of Larkspur, where a spectacular but impractical open-air terminal had been built at an additional cost of $25 million.

Multi-million dollar operating losses and other problems, including complaints that waves created by the fast-moving Larkspur ferries were causing serious damage to that town's shoreline, have beset the bridge district ferry service since its inception. But for all the controversy there is no denying that the boats provide one of the world's most agreeable methods of commuting. Passengers relax on comfortable seats and spend the 30- or 45-minute ride reading, working out of their briefcases, conversing with fellow commuters or simply enjoying the view. Some even catnap during the trip, an attractive alternative to fighting automobile traffic twice a day. Enroute to San Francisco, coffee, pastries and the *Chronicle* are available; cocktails and the *Examiner* can be enjoyed on the way home. Indeed, critics have protested that the tolls paid by bridge commuters are used to subsidize the cushy ride across the bay by leisurely cocktail commuters; bridge district officials respond that it is imperative to make the ferry—and bus—service as attractive as possible in order to lure North Bay commuters out of their cars.

The effort is working. Bridge district ferries, along with the buses and the van-pools, have been relatively successsful in keeping a lid on bridge traffic. In 1970 some 33 million vehicles crossed the bridge, an increase of more than 13 million over the 1960 traffic ten years earlier. But ten years later, in 1980, the number increased by only about 3 million, with 36.4 million vehicles crossing in fiscal 1980–81.

By the summer of 1971 more than $150 million had been collected in tolls, and on July 1 of that year the Golden Gate Bridge was officially paid for. When bridge commuters pointed out with considerable logic that the crossing should now be free, the bridge district explained—with logic of its own—that the tolls, which produced much of the revenue needed to maintain the bridge and to subsidize the bus and ferry systems, should be considered permanent.

Over the years bridge tolls have been as capricious as the winds blowing through the Gate. The 50-cent each way automobile toll of opening day lasted only seven months before it was cut in half during a price war between the bridge and the Southern Pacific Ferry Company. The "war" ended after 11 days, and the toll returned to 50 cents, where it stayed through World War II.

An efficient one-way toll was introduced in 1968: pay a round-trip toll in one direction, cross free in the other. The first of its kind in the land, it was quickly adopted by the Bay Bridge and by other toll bridges throughout the country. The 1977 toll was fixed at $1, collected only from southbound traffic. In 1981 the toll was raised to $1.25, then dropped back to $1 four months later after bridge authorities realized that handling the coins and making change slowed down toll collections and created additional traffic delays. To provide the revenues that the $1.25 toll would have earned, a unique split toll formula was devised, $2 from midnight Thursday to midnight Saturday, $1 at all other times.

The bridge toll will surely continue to increase, as will the costs of maintaining the bridge and subsidizing the bus and ferry systems. But given its rise and fall since Opening Day in 1937, it is imprudent to predict where the figure will stand one year or ten years from now. One thing is for certain, though: a free round-trip drive across the Golden Gate Bridge has become the impossible dream.

BICYCLES

WHY DO THEY JUMP

The Golden Gate Bridge is practically suicide-proof. The guard rails are five feet, six inches high and are so constructed that any persons on the pedestrian walk could not get a handhold to climb over them. The intricate telephone and patrol systems will operate so efficiently that anyone acting suspiciously would be immediately surrounded. Suicide from the bridge is neither possible nor probable.

Joseph Strauss
San Francisco Call-Bulletin, May 7, 1936
(one year before the bridge opened)

But when the guard railings were installed early in 1937, they measured only three feet, six inches high. The shorter railings gave pedestrians and motorists a better view of San Francisco Bay. They also gave pedestrians who wanted to climb over the railing an easy handhold.

The bridge had not been opened three months when on August 7, 1937, Harold Wobber, 49, a World War I veteran, walked to center-span and climbed over the pedestrian railing. He fought off a bystander who tried to restrain him, then jumped 260 feet into history, the first recorded suicide off the Golden Gate Bridge. By 1982 nearly 800 people were known to have followed. The actual number can never be known.

The beauty of the bridge is thus inseparably tied to its haunting image as a place of death. That image has made the Golden Gate Bridge the number one suicide site in the world. Other notorious attractions to suicidal individuals, such as the Eiffel Tower and the Empire State Building, have long since withdrawn from the grim competition for suicide deaths, by erecting barriers. On the Golden Gate Bridge the count continues to climb.

Complete records were not kept during the early years, but eventually bridge authorities set up a detailed filing system to keep track of "actual" suicides and "possible intents." Actual suicides include any person who is seen hitting the water below the bridge, whether or not a body is recovered. (The California Highway Patrol records a suicide only when a body is recovered.) According to bridge records, ten people jumped to their deaths in 1940; but during the war years, when young men were in combat and the nation was fully occupied with the war effort (and National Guardsmen stood patrol on the bridge), the number of suicides declined: two in 1942, three in 1943, four in 1944. In the final year of the war, 1945, the toll began to rise again: ten known suicides. In 1948 the number climbed to 19.

The Golden Gate Bridge is sometimes described as a barometer of an increasingly troubled society. During the so-called innocence of the 1950s, bridge suicides averaged ten per year. In the 1960s that figure more than tripled. The memorable "Summer of Love" in 1967 brought an estimated 200,000 young people to San Francisco from throughout the country. The following year 30 suicides were counted, a record number. Many were young people.

The number of bridge suicides per year continued to increase during the 1970s. At the same time the average age of the victims dropped significantly, from early fifties to early twenties, where it remains.

The ironworkers, who spend much of each day climbing over every inch of the bridge, are often called upon when a person has climbed over the pedestrian rail and is clinging to a steel girder a few feet below. If bridge officers and highway patrolmen are unable to talk a potential suicide off the bridge, an ironworker is summoned. But, "It's a terrible experience to have a guy within arm's reach and lose him," and every ironworker has a chilling story about his attempt to prevent a suicide. One longtimer recalls:

One day one of the electricians and I were out on the bridge. We could hear this woman screaming and we didn't know where in the hell it was coming from. We were on the sidewalk. We looked around, found her hanging on the outside of the bridge by her fingertips. We climbed over the side and each got hold of a wrist. Fortunately she was only about 90 pounds wringing wet; otherwise I don't know if we could've done it. She'd gotten that far and decided she wouldn't do it. There wasn't anything below her except water.

Many come to the bridge planning to jump but have second thoughts after they have taken the step over the pedestrian railing, and climbed down onto one of the steel beams—last stop on the way down. Retired bridge captain Ed Moore recalls one tragic case of an individual who apparently wanted to insure against a change of heart and combined two deadly forms of self-destruction:

The guy climbed over the rail and down into the superstructure. Then he cut his wrists. He thought he would collapse after loss of blood and fall off. Instead, he became wedged between the girders and didn't have enough strength to throw himself off. The next morning, in heavy fog, a soldier walking toward the Presidio heard him cry out. It was Saturday and nobody on the weekend crew would go over the rail because it was slippery with all the fog. We called the fire department, but the captain refused to send his men over the rail. We called the ironworkers. They went down and got him out. A week later he hung himself at home.

Moore estimates that during his 31 years on the bridge, "We saved 15 for every person who jumped." Many people come to the bridge and threaten to jump. "Possible intents" outnumber actual suicides by about four-to-one. In 1977, when 42 people were known to have jumped, another 151 were removed from the bridge after threatening suicide or exhibiting behavior judged by bridge officers to be possibly suicidal.

Attempted suicide is not a crime in California. A person can, however, be held for observation for up to 72 hours if he is judged to be a danger to himself or others.

Bridge officers feel frustrated and not a little angry when they prevent a suicide only to see the same person return and repeat the threat again and again. Their anger is directed at a system seemingly unable to cope with the severely disturbed. After threatening to jump, an individual is usually taken to the psychiatric ward of a nearby hospital. Too often, this is little more than a revolving door. "We've had cases where they beat us back to the bridge," says one bridge officer.

One popular but unsupported theory about bridge suicides, often labeled the "end-of-the-trail theory," holds that people who consider themselves failures move West in pursuit of new beginnings. Once in San Francisco, but still unable to come to terms with life, these people see the end of the American continent as literally the end of the line, an appropriate place to die. This romantic theory has no basis in fact. Studies have shown that nearly 80 percent of the people who jump have lived in the San Francisco area for five years or longer.

Part of the romantic myth surrounding this method of suicide is the widely held notion about the moment of death. Dr. Richard Seiden, University of California suicidologist, says, "Although jumping from the Golden Gate Bridge at a height of over 200 feet usually results in a violent disfiguring death from massive traumatic injury, these facts are not generally appreciated. Instead, the popular mythology holds that one is gently swallowed by the waves to die by drowning." A person jumping from the bridge hits the water in three to four seconds, at approximately 75 miles per hour. One study of 169 bridge suicides revealed that only eight had died from drowning; the rest died from injuries suffered on impact.

The best available studies show that jumping from the bridge is 98 percent fatal. Approximately a dozen jumpers are known to have survived, mostly by luck. Most survivors hit the water feet-first. It is, however, almost impossible for a person to control the position of his body during the 260-foot fall at 75 miles per hour. Of the six survivors inter-

Somber light from the western sky casts a shadow image upon the waters of San Francisco Bay, repeating the harmony of form and design that is the Golden Gate Bridge.

viewed by Dr. David H. Rosen, staff psychiatrist at the Langley-Porter Neurological Institute in San Francisco, four said they had blacked out before hitting the water; two who were conscious at impact said they landed feet-first.

A contributing factor to the number of Golden Gate Bridge suicides is the easy accessibility of the bridge to pedestrians. Relatively few suicides drive their own cars to the bridge; they arrive on foot, by bus or by taxi. Unlike most high suspension bridges, including the Bay Bridge, where only 152 suicides have been recorded since its opening in 1937, the Golden Gate Bridge was designed to accommodate pedestrians. Even the guard rail that Joseph Strauss predicted would prevent suicides was redesigned to improve the view from the bridge. (Another myth—that people jump from the east side of the bridge so they can catch a last glimpse of San Francisco as they plunge to their death—is dispelled by the fact that the pedestrian sidewalk on the east side is open seven days a week, while the walkway on the west, facing the sea, is open to pedestrians only on weekends.)

All of this has led to what is currently, and has been for some years, the single most controversial issue on the Golden Gate Bridge: a suicide barrier. From the day Harold Wobber jumped, suicide prevention groups have lobbied the bridge district board of directors to install a barrier. The clamor grew stronger as the suicide rate climbed during the 1960s and 1970s. Several designs for a suicide railing were submitted and the issue was extensively debated among suicide experts, the public, the press, the board of directors and politicians at various levels of power. The argument from those favoring a barrier is basic: it would save lives. The opposition is less direct. Cost is often cited; construction estimates run as high as $4 million. An almost unspoken factor in the opposition to a higher railing is the fear that any new structure would mar the architectural beauty of the bridge and, of course, ruin the view.

But the argument most often used against the suicide barrier is that those with suicidal intentions would simply find some other means of killing themselves. Experts believe that this is untrue. "The untested assumption that persons stopped from committing suicide from the Golden Gate Bridge will inexorably go to another location does not square with the facts that we know about suicidal persons and their behavior," says Dr. Seiden. "To begin with, one must appreciate the fact that suicidal persons are extremely ambivalent. That is, they wish to die but they also wish to live and to be saved. The threat of suicide and even suicide attempts themselves are seen as a cry for

help, a fight between life and death rather than a singular desire for self-destruction. Moreover, the period of extreme danger is limited. It has been estimated that the most severely suicidal person is in danger of killing himself only over a period of hours or days rather than an entire lifetime."

Dr. Rosen conducted an intensive follow-up study of seven men and one woman who had jumped from the bridge and survived. The youngest was 16; the oldest 36; the average age was 24. One was married, one divorced, the others unmarried. Three of the eight were in psychiatric treatment; one was an inpatient who had left the facility with a pass when he jumped. Dr. Rosen interviewed six of the survivors and reported his findings in 1975 in the *Western Journal of Medicine*:

All six of the interviewed Golden Gate survivors' suicide plans involved only *the Golden Gate Bridge. Four of the six said they would not have used any other method of suicide if the Golden Gate Bridge had not been available (for instance, if there had been a suicide barrier).*

One survivor associated the beauty of the bridge with death and jumped from it because "I was attracted to the bridge—an affinity between me, the Golden Gate Bridge and death—there is a kind of form to it, a certain grace and beauty. The Golden Gate Bridge is readily available and it is connected with suicide."

Another survivor said jumping from the Golden Gate Bridge was "a romantic thing to do," and that it assured "certain death in a painless way." A different survivor described suicide from the bridge as "easy" and a "sure thing." Another said the Golden Gate is "the only sure way." The remaining survivor said the extensive press coverage and the "notorious fame" that surrounded Golden Gate Bridge suicides, plus its "easy accessibility," made it the "only method" of suicide for him.

According to Dr. Rosen, two of the six interviewed made subsequent suicide attempts, both by drug overdose. All six favored the construction of a suicide barrier and felt it would save lives. Concluded Dr. Rosen:

For all of them, this bridge had a special and unique meaning. Often this was related in a symbolic way to the association of the Golden Gate Bridge with death, grace and beauty. The fact that the Golden Gate Bridge leads the world as a location for suicides should be knowledge enough for us to begin to deromanticize *suicide, especially as it relates to the Golden Gate Bridge, but also in a general way.*

In addition to deromanticizing *suicide and death, especially as they relate to the Golden Gate Bridge, these findings point to a need to do something practical in order to prevent further suicides from that structure.*

I underscore and concur wholeheartedly with the survivors' unanimous recommendation that a suicide barrier should be constructed on the Golden Gate Bridge.

One of the strongest arguments in support of a barrier is represented by its success in ending death at other structures which were once notorious suicide attractions. For many years, the one site in the world that ranked with the Golden Gate Bridge in the number of suicides was the Eiffel Tower. Between 1889 and 1968 more than 350 people jumped to their deaths from the tower; suicides stopped altogether after barriers were built. *San Francisco Chronicle* columnist Herb Caen reported the rather peculiar manner in which the Eiffel Tower withdrew from the world suicide standings:

When the Gate Bridge achieved No. 410, I phoned the French Consulate to check on the Eiffel's record and got such evasive answers . . . I put our man in Paris, Ferris Hartman, on the job and he came up with surprising information. The Eiffel's top figure, he found, was 352, scored in April 1968—at which time the Gate Bridge's suicide figure was also 352! Apparently the French sensed defeat, for when, on June 27, 1968, one Jean Tapie Carraze jumped to his death from the Eiffel Tower because he had lost his driver's license (how Gallic), he too was listed as No. 352. The next suicide was numbered 351 and the numbers have been decreasing ever since.

"What is the latest figure?" Mr. Hartmann asked the police chief . . . near where the Eiffel is located. "Around 300," he was told. "Since we put up barriers, no more suicides." Conclusion: in this melancholy category, the Gate Bridge now has an insurmountable lead. Unless they put up barriers and start counting backwards, too.

Other examples of effective suicide barriers are found closer to home. The Arroyo Seco Bridge, built in 1913, is an overland highway bridge spanning a gulley near the Rose Bowl in Pasadena. By 1936, 80 people had leaped to their deaths from that bridge. Public pressure resulted in construction of a suicide barrier. Since then, there has been only one suicide from the Arroyo Seco. There were 16 suicides by people who jumped from the popular observation tower of the Empire State Building in New York City between 1931 and 1947, the majority of them after World War II. The sudden rash of suicides resulted in the construction of a tower fence that curves in at the top. It has prevented any further suicides from the Empire State Building.

Few people are aware that there is already a precedent for a suicide barrier on the Golden Gate Bridge. When the bridge was designed, a glorious steel arch over Fort Point was included to preserve the old red-brick fortress. But a few people jumped from the bridge and were killed when they landed on top of the fort. Pedestrians also had the regrettable habit of throwing litter over the side and into the fort below. In 1970 Fort Point was declared a National Historic Site and opened to public tours. To spare the tourists from flying missiles and from witnessing suicide leaps, bridge authorities installed an unsightly nine-foot chain-link fence along the railing above Fort Point.

Still, the call for a suicide barrier across the full length of the bridge continues to go unanswered. Dr. Seiden argues that the evidence is overwhelming justification for a barrier. "There is little question," he writes, "that the Golden Gate Bridge has become a suicide landmark with a fatal mystique that draws the suicide-prone. Previous studies of suicide landmarks indicate that physical barriers will dramatically reduce suicide deaths at particular locations. There is overwhelming evidence that the ambivalence and acuteness of suicidal behavior lead to a favorable prognosis for people who are stopped or apprehended from completing their suicides. There is little or no evidence to indicate that they will invariably go someplace else to kill themselves."

Or as one of the nearly 800 *known* suicides asked in the note he left behind when he leaped to his death from the Golden Gate Bridge, "Why do you make it so easy?"

Joseph Strauss believed suicide from the bridge was "neither possible nor probable." Unfortunately, the first victim jumped in August 1937, a short three months after Opening Day. Since then, nearly 800 are known to have followed.

The view from the bridge is one of its assets, to many its greatest. It is a widely-held but limited notion that this outstanding view would be impaired by a suicide barrier. Talented industrial designers could easily create an aesthetic, harmonious railing that would protect both the view and the viewer. The retouched photograph on the right illustrates only one such possible solution.

Photo © Mush Emmons

Seldom has the Golden Gate Bridge been closed to all traffic. The first occurrence, in December 1951, was due to gale force winds that shook the towers and twisted the roadway. It was closed again in the winters of both 1982 and 1983, when continuous storms drenched the Bay Area.

From the top of the tower to the roadway is a distance of nearly 500 feet; from the roadway to the water is well over 200 feet. The fluted brackets inside the upper corners of the portals are the only decorative touches allowed in an already pragmatic, harmonious design.

FORT POINT

In early drawings for the Golden Gate Bridge the massive concrete block of the south cable anchorage was designed to rest on a level spit of land that is the northernmost point of the San Francisco Peninsula. Had these designs been followed, it would have meant the destruction of one of the most historically significant structures in San Francisco: Fort Point.

Given his respect for structures—even the cold form of military architecture—it is doubtful that Joseph Strauss ever seriously considered demolishing the 1861 fort.

So plans were ordered drawn to preserve Fort Point by building around and above it. This meant locating the cable anchorage behind the fort and constructing a steel arch over its three stories, a detour that cost considerable time and money. As a result, not only was Fort Point saved, but its setting was dramatically enchanced: a jewel of a nineteenth-century brick fortress nestled beneath a steel engineering marvel of the twentieth century.

Strauss's first attempt at supporting the roadway over Fort Point was an elaborate design calling for a special cable suspended beneath the bridge deck. Chief assistant engineer Clifford Paine intervened here—as he had on several other of Strauss's design flourishes—with a counterproposal. Paine argued successfully that the fort would best be spanned by a conventional steel arch.

During bridge construction, engineers and other key personnel set up offices inside the fort. Also within its walls was a cafeteria for the bridge workers, while the center courtyard became a temporary parking lot.

Not only was the fort itself saved, an historic seawall was preserved as well. Before widening and paving the site and building a new breakwater, workers dismantled the granite seawall, removing hundreds of tons of material piece by piece, storing the pieces for four years, then rebuilding the seawall with the original granite.

Though the Spanish established a mission and presidio in San Francisco in 1776, it was not until 1794 that they built a fort on the site overlooking the entrance to San Francisco Bay. Named Castillo de San Joaquin, it was poorly constructed of adobe brick and assigned the task of protecting the Spanish presidio, itself located more than a mile distant.

An inspection by the Spanish of the adobe fort in 1796 did not inspire confidence in the might of the new harbor guardian: "The structure rested on sand and decaying rock; the brick-faced adobe walls crumbled at the shock whenever a salute was fired; the guns were badly mounted and for the most part worn out, only two of the thirteen 24-pounders [referring to the 24-pound cannonball] being serviceable or capable of sending a ball across the entrance of the port. The whole work, protected by an adobe wall with one gate, was commanded by a hill in the rear, and the garrison of a corporal and six artillerymen was altogether insufficient."

While the frail little fort was fortunately never challenged by enemy guns, it was under constant attack from the elements. Years of pounding rains and high winds, plus the occasional earthquake, had reduced Castillo de San Joaquin to a shambles by 1835, when the Spanish abandoned the fort and moved most of their presidio garrison to Sonoma.

The undefended southern flank of the entrance to San Francisco Bay lay ignored until 1846, when the United States began its conquest of California. With the discovery of gold in the northern California foothills two years later, San Francisco suddenly became a rich port, and harbor defense took on new significance. Seacoast cannons and howitzers were mounted at the point and plans were made for repairing and arming the old Spanish fort. Meanwhile, Army engineers began work on a grander scheme for the defense of San Francisco Bay.

In 1853 Congress appropriated $500,000 to start construction

The 50,000 ton south cable anchorage was constructed adjacent to Fort Point. The roadway would eventually curve above and around the old brick fortress, now part of the Golden Gate National Recreation Area. The barbette tier of gun emplacements is clearly visible on top of the fort.

Not only was Fort Point saved, its setting was dramatically enhanced: a jewel of a nineteenth-century brick fortress nestled beneath a steel engineering marvel of the twentieth century.

of a new fort on the old Castillo site, designated Fort Point. The money was also to be used to erect a fort on Alcatraz Island and to build gun batteries on both Angel Island and Point San Jose (now Fort Mason), about two miles east of Fort Point.

Fort Point itself was completed in 1861 at a cost of $2.8 million. Construction of a 2,000-foot seawall cost another $400,000; it was not finished until 1870. More than 120 years later, Fort Point stands intact, a fine example of the multiple-tiered brick forts built by the Army in the last years before the Civil War. Similar in design to Fort Sumter in South Carolina, the fort is built in an irregular quadrangle with seven-foot-thick walls rising 45 feet high and enclosing a paved courtyard. On the three sides facing the water are three tiers of brick arches supporting gun ports. A fourth tier of guns, the barbette tier, completely encircled the top of the fort. At one time Fort Point could boast of 126 cannon positions for defense against land or sea attack.

During the Civil War, Fort Point braced itself for threatened raids by Confederate privateers on the gold shipments. At the time the fort's armament included approximately 55 cannons, from 24- and 42-pounders to 10-inch Columbiads. But the Civil War ended without a shot being fired at or from Fort Point.

Development of breech loading guns and other advances in weaponry in the late nineteenth century rendered Fort Point's seacoast guns obsolete. In 1897 many of the original cannons were hauled off to be displayed as historical pieces at other Army posts; two of these guns can be found today not far from the fort, on the parade grounds of the Presidio of San Francisco. In 1901, what remained of Fort Point's coastal defense weapons was sold for scrap.

Army garrisons came and went at Fort Point during the years following the Civil War. The 66th Company of the Coast Artillery was quartered there when San Francisco was nearly destroyed by earthquake and fire in 1906. The big earthquake shook the thick brick walls of the fort, moving them as much as eight inches, but no major structural damage was sustained. Nonetheless, the 66th Artillery was promptly transferred to new quarters.

In 1913 there were plans to transfer the Army Detention Barracks from Alcatraz Island to Fort Point. The interior tiers of the fort were converted into medium security dormitories but the government changed its mind and abandoned the project before a single prisoner spent a single night in the fort. The same areas that had been remodeled to detain military prisoners were eventually used as officers' quarters during World War I and as bachelor officers' quarters after the Armistice.

Troops returned to Fort Point at the outbreak of World War II, when there was widespread fear that the Japanese were planning to follow their attack on Pearl Harbor with a strike at the coast of California. Two three-inch rapid-firing guns were mounted atop the fort walls and aimed toward the channel. Two searchlights were operated from the top of the fort to illuminate the entrance to the bay. Mobile antiaircraft guns were positioned in front of the fort, facing the bay. Men of the Sixth U.S. Coast Artillery remained at Fort Point until 1943, when the tides of war in the Pacific had turned and the U.S. was no longer fearful of a Japanese attack upon the California coast. The removal of the World War II troops ended Fort Point's 82 years of active service on behalf of the U.S. military.

The construction of the Golden Gate Bridge was the closest the old fort has ever come to being assaulted by anything other than the forces of nature. Fort Point has never been fired upon and its cannons have never fired a shot in anger.

In 1959 the Fort Point Museum Association was formed to preserve the structure. Years of hard work finally resulted in Fort Point's designation as a National Historic Site in 1970. One year later it was dedicated as the first National Park site in the San Francisco Bay Area and it subsequently became part of the vast Golden Gate National Recreation Area. In 1982 more than a million people visited the fort, admiring and photographing the exceptional views from below the Golden Gate Bridge.

Although Fort Point has survived the ravages of time and weather remarkably well, restoration by the National Park Service has enhanced its appeal as an historic landmark. The fort exists today much as it appeared between 1861 and 1913 when it was fully operational.

On top of the structure is a lighthouse and two seacoast cannons. The cannons, an eight-inch Columbiad and a 32-pounder, are reproductions donated by the Fort Point Museum Association. The lighthouse, built in 1864, was originally outfitted with an oil-burning lamp. It was maintained by lighthouse keepers who lived on the bluff above the fort and walked to work on a private footbridge strung between the lighthouse and their clifftop quarters. The beam was a beacon for ships sailing through the Golden Gate, but because the superstructure of the new bridge blocked out much of the light, it was extinguished in 1935.

While the emphasis at Fort Point is on its long military history, the presence of the world's most beautiful bridge arching gracefully above has added immeasurably to its standing among San Francisco landmarks. Park rangers include some of the bridge history during the regu-

Fort Point looks much the same today as it did in 1907. The long seawall was also saved when Strauss chose to preserve the fort itself. Dismantled piece by piece, the original granite seawall was carefully reassembled when bridge construction was complete.

larly scheduled public tours of the fort, and once a day visitors are treated to a documentary film made by Bethlehem Steel as the bridge was being built. This intriguing film, now nearly half a century old, shows the Golden Gate Bridge rising piece by piece, slowly joining San Francisco to Marin County. When the last sections of the roadway are finally connected, the audience often responds spontaneously with applause and cheers, and the brick walls of the old fort reverberate once again in celebration of the historic spanning of the Golden Gate.

Fort Winfield Scott—the official name of Fort Point—housed the offices of bridge engineers and other key personnel. Its courtyard became a temporary parking lot, and within its walls a cafeteria fed hungry construction workers.

JOB WITH A VIEW

In its first year the Golden Gate Bridge cost less than a half-million dollars to operate and maintain. Salaries were not much higher than the $4 to $11 per day that had been paid to workers during the bridge's construction. Today the Golden Gate Bridge, Highway and Transportation District has nearly 850 full-time employees and total operating and maintenance expenses of more than $40 million (1981 revenues exceeded $43 million). Salaries range from $10.75 an hour for toll collectors to nearly $16 an hour for ironworkers and $17.60 for painters. But the bridge is now only a part of the district's far-reaching transit system. There were 167 men and women working on the bridge in 1982, while nearly 500 were needed to keep the buses running and another 73 were employed in the ferryboat division. The Golden Gate Bridge not only serves as the vital link between the North Bay counties, but the tolls collected on the bridge support the only mass transit system between San Francisco and Marin and Sonoma counties.

"People from all over the world come to study the Golden Gate Bridge as a multi-mobile transit system," says one bridge administrator, who also points out that the system is unique among public works projects in the United States. "The bridge was built without tax revenues and we still have no taxing authority. Yet we have a fully in-place transit system."

While the district's expansion into other forms of mass transit has brought some needed relief to congested bridge traffic, the world of the bridge district still revolves around its bridge. Painters work almost continuously to keep the bridge looking beautiful and to protect it from corrosion and rust. It is a job that started almost as soon as the construction of the bridge ended, and it is a job without end. A small crew of painters who had worked during the construction period were hired at the end of 1937. Starting at the bases of the towers where rust had already attacked rivet heads and surfaces, they removed the rust and applied a protective coating of coal tar paint. Corrosion of the bridge steel by the salt-laden fog was even more severe than engineers had anticipated, and so the original painting and maintenance program had to be accelerated.

A full crew of ironworkers also keeps busy on the bridge replacing corroded steel, assisting with the complicated rigging systems used by the painters and performing a variety of maintenance jobs to protect the bridge from the constant assault by weather, traffic and age. Inspectors visit the bridge several times a week, examining the cables and checking for signs of deterioration in the steel. The bridge district also employs machinists, operating engineers and road maintenance crews. A vault keeper presides over the toll money as it pours in. To accommodate the changing flow of commuter and holiday traffic, lane workers go out in trucks three times a day and move the rubber "paddles" which control the traffic: 3 lanes each north and south during normal hours, 4 lanes for commuters during rush hours. The bridge district maintains a permanent fleet of service vehicles, including a fire engine which escorts trucks carrying inflammable materials across the bridge. Flat tires, mechanical failures and automobile accidents keep service crews busy. Unlike most toll bridges, the Golden Gate does not fine unlucky motorists who must be towed off the bridge with a flat tire, dead engine or empty gas tank. A minimum amount of gasoline is sold to drivers who choose the Golden Gate as a memorable place to run out of gas. A crew of electricians maintains the bridge's complex electrical system which includes the sodium and mercury vapor lamps that light the roadway, the airplane beacon lights atop the towers, the computerized machines in the toll booths that record the passage of every vehicle crossing the bridge, and the all-important foghorns. And, of course, there is the person with whom virtually every southbound driver must do business, the toll collector.

THE TOLL COLLECTOR

He stands inside a booth the size of a broom closet, sticks out a hand as each car approaches, and takes the driver's money. He punches a button, makes change if necessary, and says "thank you," car after car, hour after hour. It would be among the easiest and most boring of all jobs, except for one fact: toll collectors aren't dealing with cars, they are dealing with people. And people do the craziest things on the Golden Gate Bridge.

When a toll collector puts a hand out, he cannot assume that money will be offered. Toll collectors have been handed loaded guns, dead fish, live kittens, assorted garbage, boxes of bees, pizza, salamis, fruit and every other food imaginable. They have reached out to collect tolls heated red-hot with cigarette lighters or attached to electrical shock devices. Drivers have been known to carry their toll money in their teeth, behind their ears, between their toes and in wooden spoons. Toll collectors have been caught in the middle of fist fights, lovers' quarrels, robberies, prison escapes and drunken parties. They have been bitten by dogs, stung by bees and propositioned by drivers. They have had their arms twisted, been spit at, and had money and verbal abuse thrown at them on a regular basis. Toll collectors have also been called upon to round up runaway cattle, wild deer, skunks, raccoons, dogs and cats. They must deal with women who back up traffic while fixing their make-up, drunks who fall asleep at the wheel, lost tourists, toll evaders, daydreamers who forget to pay, drivers who forget they are broke until they reach the toll booth, and drivers who argue they shouldn't have to pay because the bridge has already been paid for.

A monotonous, boring job? Sometimes, perhaps. But not when sex acts are being performed in passing cars, when a car full of naked young women drives through, when a well-endowed woman drops the toll at her feet and asks the collector to please retrieve it, or when the window rolls down to reveal the barrel of a shotgun.

It is possible that no one in the world has more person-to-person contact with more people on a daily basis than a Golden Gate Bridge toll collector. Individually they probably take more tolls from more cars than the toll collectors on any bridge in the world. One toll collector can handle 700 cars an hour during the morning commute traffic, and by the end of a shift he may have taken tolls from 4,000 people. Dealing one-to-one with that many people every day, they are bound to witness the complete spectrum of human behavior. And the climate for eccentricity is strong here. For some people the Golden Gate—the most famous bridge in the world—is a siren call to perform.

Conflict and comedy aside, the toll collector's job is that of a high-pressure cashier to the fast-moving world of the American motorist. A toll collector needs the basic instincts of a bank teller, a traffic cop and a public relations person. Some put it all together better than others, but the consensus of Bay Area drivers is that the toll collectors on the Golden Gate Bridge are the best in the business. They maintain a high degree of efficiency while handling thousands of dollars, and keep a happy face doing it. Efficiency is the easier of the two objectives.

The stresses of the job are described by former toll collector Antonio Santa Elena. Santa Elena, who spent 12 years as a collector on the Golden Gate Bridge, says, "Few people appreciate the long hours a collector works, exposed to inclement weather and inhaling the fumes from car exhaust, attuned to the constant nerve-shattering noise of running motors, loathed and derided at times by impolite drivers, and often bearing the brunt of criticism against the bridge district. What makes a collector lose his smile and his 'thank-you'? Rude, ill-mannered and mean drivers."

There are good experiences too. Many people crossing the

There is no grander view of the bridge and its magnificent setting than from the hills of Marin in the Golden Gate National Recreation Area. When the night sky is clear and the moon full, awe-struck crowds gather to gaze down upon the breathtaking panorama.

The fireboat *Phoenix* welcomes visitors through the Golden Gate with a unique greeting of 50-foot plumes of spray from its deck guns. Capable of pumping 9600 gallons per minute, the *Phoenix* is an important component in the fire-fighting arsenal of the San Francisco Fire Department.

bridge for the first time still feel compelled to congratulate the collectors ("You have the most beautiful bridge in the world"). Warm relationships can develop between collectors and regular commuters who sometimes bring gifts or exchange a quick bit of gossip as they roll through the toll lane every morning. Collectors are often the first to know what fruits are coming into season when drivers start handing them paper bags full of fresh pickings from the fertile orchards north of the bridge.

However, neighborly exchanges between toll collectors and motorists have diminished over the years, while abuse has escalated. Traffic has increased, tolls have risen, the bridge district has become more controversial and life in general seems to be more frustrating and less friendly. Drivers who are mad at the bridge district—or just mad, period—are prone to take it out on the toll collectors. "I thought this damn bridge had been paid for," is a common gripe. Most of the abuse is verbal or petty: dropping the toll money out the window, slapping the toll collector's hand with it, trying to underpay, or paying with pennies. But on occasion the abuse is more serious. In late 1981 a collector had her arm severely wrenched and missed several weeks of work after a driver grabbed her hand, twisted her arm and slammed it up against the car door, then sped away.

Toll collectors are supposed to keep their personal feelings to themselves no matter what the provocation. But sometimes those feelings boil over. One Christmas Eve, shortly before midnight, a car stopped at the toll booth and the driver struggled to push a large box through the window. "For you and your friends on the bridge, Merry Christmas," he said, and drove away. It was a case of bourbon. A bottle was dispensed to each of the collectors on duty and they stashed them away in their lockers. All except one collector who decided to enjoy some Christmas spirit on the spot. Before long he was out of his booth and standing in the middle of the traffic lane, waving cars aside and calling their drivers every name he had been called during his years inside the toll booth. Bridge officers finally led him away and sent him home. Because of his long record of good service, his only punishment was a brief suspension.

Golden Gate Bridge toll collectors are officially called bridge officers. This is more than a polite title. Occasionally, for example, they may be called upon to watch for a car carrying an armed robbery suspect.

Retired bridge captain Ed Moore likes to tell about the less threatening emergency alerts on the Golden Gate Bridge. There was the San Francisco baker who called to ask that bridge officers intercept his driver, who had left 1500 loaves of bread behind. A woman once called from San Jose and demanded that her son-in-law be halted and sent back because "he forgot something—me!" One of Moore's favorite stories concerns a man who was driving his pregnant wife across the bridge to a San Francisco hospital. Traffic had been held up by a bomb threat, but the baby couldn't wait. An announcement over the bridge's public address system summoned a registered nurse, and a baby girl was delivered in the car as it sat stalled in the traffic. The grateful parents named their baby "Golden," and when Ed Moore recalls the incident, he is fond of pointing out that "it's a good thing she wasn't born on the Dumbarton Bridge."

Where two or more automobiles are gathered there is always the chance of an accident, or an incident. Drivers who cut in front of one another crossing the bridge or approaching the toll booths sometimes end up in fist fights when they stop at the toll plaza. Collectors must also now contend with toll evaders, a problem that increased dramatically from perhaps once a month in the early 1960s to a dozen or more a day by 1982. A collector can try to write down the license number of a car as it speeds past the toll booth, but the chances that the courts will prosecute a driver for evading a $1 toll are almost nil. One notorious tollgate crasher, a doctor who lived just over the bridge in Mill Valley, was eventually taken to court by the bridge district. He paid a small fine and went right back to gate-crashing.

Then there are the "no-funds" or "brokers"—drivers who, by design or by accident, don't have the money to pay the toll. These drivers are directed to an office adjacent to the toll plaza where they are asked to sign a promissory note to pay the toll, and may be required to leave the bridge some collateral. Through the years they have left jewelry, radios, calculators, musical instruments, clothing, toupees, wigs, dolls, rosaries and prayer books. One man left his artificial arm. Another left his upper dentures. And there is the occasional driver who wants to leave a spouse as collateral; at least once, a man did just that.

Anything can happen on the bridge at any hour, but some of the strangest incidents occur between midnight and dawn. Santa Elena witnessed one event that was out of the ordinary even by Golden Gate Bridge standards. It was a foggy night about 3am. A young go-go dancer had finished her shift and was heading home across the bridge. With little traffic at that hour she had her choice of toll lanes. She had more than a casual acquaintance with one of the collectors, so she drove up to his lane, stopped, got out of her car and stepped quickly into the tiny toll booth. The lights in the booth blinked off, and while this is one record that cannot be confirmed, in all probability the go-go dancer and her friend in the toll booth were responsible for another "first" on the Golden Gate Bridge that night.

Collectors work eight-hour shifts and are given at least three half-hour "blows," or breaks from the pressures of the booth. During periods of unusually heavy traffic, breaks may last up to an hour, to rest weary feet and steady frazzled nerves.

Each collector has a personal system for efficiently handling the thousands of transactions per shift. As each driver pays the toll, the collector pushes a button which automatically records the vehicle type and amount of toll. "Take the money, hit the button; take the money, hit the button." But various special tolls, including commute tickets, complicate the system. In the truck lane, the toll fee must be determined individually by the number of axles on the particular truck. Collectors keep "setups" of specific amounts of cash so they can immediately make change for large bills without delaying traffic. They must also be alert for drivers who try to deceive them into thinking that change is needed for a large bill. "Sorry, sir, but a twenty is the smallest I've got." If a collector makes change before he sees the bill come through the window, the driver may speed off with $19 in change from a $1 bill.

Precise daily records are kept on each collector's ratio of mistakes. Those at the top of the performance list may be off by only a nickel for every thousand dollars in tolls collected. Those at the bottom might be off by a dollar or more for every thousand dollars. The error-prone collectors are expected to improve their "system" or start looking for another line of work. But few are dismissed. More often they leave by choice, no longer willing to tolerate the abuse that comes with the job.

Backing up the toll collectors are the sergeants and lieutenants who operate from a large office at the west end of the toll plaza. Inside is the 24-hour nerve center of the Golden Gate Bridge—a control console resembling the instrument panel of a 747. This system enables officers to monitor vehicular and pedestrian traffic, and to take prompt action when problems occur. Some problems transcend the resources ordinarily available to bridge officers.

- *Two young motorists are throwing beer cans at each other as they drive across the bridge. At the toll plaza the driver of the first car jumps out, runs back and starts beating on the hood of the second car with a lead pipe. The driver of the second car gets out and pulls a gun.*
- *Girl meets boy in a Sausalito bar and accepts his gracious*

offer of a ride back to San Francisco. He becomes altogether too friendly in the car and when he slows down at the toll plaza she jumps out screaming for help.

- *A toll collector looks up to see several calves charging toward him. A cowboy was on his way to a rodeo at the San Francisco Cow Palace when the tailgate on the trailer flew open.*
- *A photographer, inspired by the view from the Golden Gate, starts setting up his tripod in the the center lane of the bridge.*
- *A car stops at the toll plaza, the man pays his toll, opens the passenger door, shoves his wife out, and speeds away. This is sometimes reversed; she kicks him out and drives away.*
- *A drunk driver stops at the toll booth, then passes out at the wheel. The car is locked, and even beating on the window fails to rouse him.*
- *A woman has just shot her boyfriend. He is in the car bleeding. She hands the gun to the toll collector, who thinks it's a joke and accidentally fires a shot into the floor of the toll booth.*

At a time when the potential for trouble on the bridge has increased — more traffic, more vandalism, more confrontations between motorists, more danger of violence—backup support for bridge officers has weakened.

The California Highway Patrol (CHP) once maintained an office and a holding cell near the toll plaza at the south end of the bridge, but it was closed several years ago. Now the CHP merely keeps a telephone and a desk in the bridge captain's office.

In any event, the powers of the CHP on the bridge are confined to vehicle-related incidents. This sometimes puts bridge officers, whose police powers are strictly limited, at a disadvantage. A typical example occurred one afternoon when two tourists were photographing the south tower and a young man standing nearby objected to being included in the pictures. He assaulted the tourists and was beating them before bridge officers intervened. Bridge officers can detain lawbreakers, but they have no powers of arrest, so the CHP was called. "Not our problem," the CHP responded. "No vehicle involved." The CHP dispatcher suggested that the San Francisco Police be called. But bridge officers call on the SFPD only when absolutely necessary. "SFPD is reluctant to come out to the bridge," says one. "They figure we should take care of our own problems out here, or use the CHP. But the CHP doesn't have the manpower they used to, so unless a vehicle is involved, they're not supposed to respond."

All bridge officers, including toll collectors, used to be armed. But several years ago the collectors stopped carrying guns after a change in status reduced their official police powers. Now

It is difficult to imagine the Golden Gate without the bridge. Aerial views make it all look so simple and natural, as if somehow the span has existed forever.

Joseph Strauss warned that "no span of steel will tolerate. . .neglect. But if it is serviced by the generations who use it and is spared man-made hazards, such as war, it should have life without end." In 1934 the first painters were sent out to protect the bridge steel from the corrosive salt air. The job has continued ever since.

only the sergeants, lieutenants and the captain are issued weapons. Officers remember only two occasions when one of them had to fire his gun. Ed Moore used his to eliminate a skunk, and Captain Lufrano's was used to shoot a deer which had been hit by a car. Despite the fact that more than $50,000 in cash is collected daily at the toll plaza, there had not been an attempted robbery on the Golden Gate Bridge until July 8, 1982. On that day a man driving a stolen car stopped at the toll booth, pointed a gun at the toll taker and ordered her to throw out her money bag, threatening to kill her if she sounded an alarm. He successfully escaped with the money—$20.

It is a commentary on the times that a few months before the first robbery in the bridge's history, those officers, who are still armed, had begun undergoing more extensive training in the use of guns.

THE PAINTERS

Nobody knows their names, yet their work is seen and appreciated by millions of people every year. From a distance one might imagine that the 31 men who paint the Golden Gate Bridge are not working on a bridge at all, but adding brushstrokes to a lifesized canvas—the ever-changing seascape of the Golden Gate.

Yet no work on canvas ever required the daring exhibited by these painters. They are men who shun the confines of a conventional working environment for the soaring beauty of the Golden Gate Bridge. Bridge painter Ali Robert Euan once explained that he chose his work because "I wanted to climb as high as I could and you can't climb much higher than this without a mountain."

If the incomparable setting and the high steel give the job its glamour, the weather turns it into a raw challenge. Yet bridge painters take it as the nature of their profession to dangle from steel cables more than 400 feet above a churning sea, buffeted by freezing winds and shrouded in fog that makes every handhold slick and treacherous. The severity of the climate is a hardship made more difficult by its unpredictability; the narrow entrance to the bay produces all kinds of weather.

Painters are issued leather gloves and rain gear. Most of them prepare for the wind and fog with coveralls worn over several layers of clothing. Choosing the correct number of layers is a daily guessing game played against the extremes of weather that may be encountered between 7:30 in the morning and 3:30 in the afternoon.

Ross Salazar, a painting superintendent who has worked for 30 years as a painter on the bridge, recalls how he would bundle up whenever he was painting the steel beneath the bridge deck that forms the arch across Fort Point. "I used to wear thermal underwear, a pair of pants on top of that, and overalls on top of that—*and* a leather jacket. And I was still cold." For the most part, men prefer working on the Marin tower and the north end of the span because the climate is consistently calmer, dryer and warmer than on the San Francisco side.

There is little seasonal relief. San Francisco draws rain in the winter, fog in the summer. And the wind never stops blowing. Only its velocity changes, from brisk bay and ocean breezes to 50 mile-per-hour storms. On many days the weather is simply too foul to send the painters onto the bridge, unless they can work in one of the few sheltered places. But even the relative comfort of the tower cells brings little relief: "You're inside the tower and you hear that big howling and you can't even hear yourself think." Some 90 days of work are lost to the weather each year.

The men curse the weather but they love the work, and the locations that appear most dangerous—the towers, for example—are often preferred by the painters. Says Salazar, "Some prefer to work up high; I would. You're out in the open. Those scaffolds are too confined. You do the same thing over and over.

The orange span in the incomparable setting is only a work-station to the painters and ironworkers who maintain the Golden Gate Bridge. Suspended high above the roadway or the waters of the bay, they move about the structure with professional ease, inspecting, sandblasting, painting and repairing.

An official Bay Area mascot, the aircraft carrier Coral Sea returns to port. Seamen at attention spell out "San Francisco's Own" on the flight deck. For 20 years, from 1963 to 1983, the Coral Sea—the oldest carrier in the US Navy—was based in Oakland at the Alameda Naval Air Station.

Do one bay (a section of bridge steel), move up to the next. I mean every rivet, every nut, every lacing, everything. It's all the same. But the towers, man, you can finish one block and move around. It breaks up the monotony."

In the mid-1960s a new painting technique was instituted. A key element in the process is the sandblasting. Clifford Paine—a man of strong convictions—was convinced that sandblasting posed a threat to the structural stability of the bridge. For 30 years his ban on sandblasting was upheld, and during that period the surface was maintained by a tedious chip-and-paint method. Air-driven chipping guns were used to remove the rust, a prime coat was applied and new orange paint was brushed on by hand. The method was slow and ineffective against the corrosive atmosphere of the Gate.

"The chipping did more harm than good," says Salazar. "They used chisels that cut into the steel. And the red lead paint wouldn't penetrate completely, leaving tiny spots where rust would quickly set in. Rust is like cancer. One little pinpoint and before your know it, it has spread in an ever-widening circle."

Soon after Paine's involvement with the bridge ended, his ban on sandblasting was lifted by the engineers who succeeded him. Since the mid-1960s the biggest maintenance job has

43

been the sandblasting, which occupies far more of the painters' time than the painting itself. Wielding nozzles loaded with 100 pounds of air pressure, painters first sandblast the steel to remove all traces of old paint and rust. A protective undercoating of inorganic gray zinc primer is then sprayed on and left to cure for 90 days. (Until the new paint job is completed, patches of the still-curing gray primer mar the color scheme of international orange. Commercial photographers who use the bridge as a backdrop must retouch their photographs that are spoiled by the effect of the bridge's temporarily splotchy paint job.)

After the primer has cured for the required 90 days, an olive-green acid-wash "tie" coat is applied as a bonding agent. Finally, two undercoats of international orange vinyl paint are applied. While the new process takes considerably longer than the old method, once the job is completed (about 1985) the bridge will require only spot sandblasting and repainting.

The new way of painting the bridge is not the only change that has affected the painters' job. Improvements in equipment have made individual tasks less tedious. Where they once hand-cranked their scaffolds up and down the towers, the men now enjoy an effortless ride on the "skyclimber," an air-powered scaffold. The spray gun is faster but more dangerous than its predecessor, the paint brush. The high-pressure guns, powerful enough to push paint through a man's skin, must be used with caution. And painters are required to wear respirators when spraying, to protect themselves from toxic paint fumes.

At one time the paint itself presented a serious health hazard. Until the advent of lead-free paint in the late 1960s, painters underwent periodic blood tests for lead poisoning. If the level in a painter's blood was found to be toxic, he was taken off the job until his lead count dropped.

Famous in part for its rigorous safety procedures, the Golden Gate Bridge has earned a commendable, if imperfect, safety record. Since it first opened for operation, two painters have died in bridge accidents. In 1967 Lee Patrick fell from a scaffold near the Marin tower. When his safety line broke he plunged into the channel. And in 1972 Andy Caton fell from a scaffold above the Marin shoreline.

THE IRONWORKERS

Ironworkers are the skywalkers of the Golden Gate Bridge. They move with ease up and down the cables, and can find their way into the most hidden corners of the steel structure. An ironworker's natural perspective is from above, looking down upon the rest of the world.

On the bridge, one of the ironworkers' primary duties is to erect scaffolding for the painters —on the towers and under the roadway. Every inch of the span must be regularly inspected, sandblasted and painted. Ironworkers also build the specialized equipment needed by the painters: trailers to haul sand to the paint crews and the large steel boxes that store the sand used for sandblasting. Even the bridge electricians rely upon the ironworkers' skills; when one of the huge electric motors must be moved on or off a bridge scaffold, the ironworkers are there to perform the heavy-duty task.

A constant vigil is maintained against deterioration of the steel span. Rivets are regularly inspected and corroded ones replaced by the ironworkers. The steel ladders used by ironworkers and painters to reach every conceivable piece of the bridge are regularly checked for dangerous signs of deterioration; some are repaired, others are replaced with new ones fabricated by the ironworkers.

Building replacement parts represents a major portion of the ironworkers' responsibilities. Ironworkers do all the repair work for the bus shelters used by the Golden Gate Transit bus system. They build and remodel bridge toll booths. They are called upon for specialized ironwork by the bridge district's auto mechanics and bodywork crews, and even do regular repair work on the ferryboats and the boat docks. Ironworkers fashioned the special equipment needed for hook-up between the ferryboats and the docks, and made alterations on the boats themselves to improve operating efficiency and passenger comfort.

Much of their work is done in the maintenance shop. When the district bought a new fire truck, the ironworkers helped to design and fabricate the unique equipment installed on the truck for its specialized use on the bridge. The ironworkers remodeled a building to provide permanent housing for the new fire truck. "Everything that anybody else on the bridge doesn't want to do, we do it," says Dick Seward, superintendent of the ironworkers.

And if something has to be done from a precarious position above or below the bridge deck, odds are that it will be done by an ironworker. Seward, a second-generation Golden Gate Bridge ironworker whose father helped to build the bridge, once spent a terrifying moment at the top of the south tower:

The worst thing I ever experienced out here was an earthquake. The initial shock happened right at noontime. I was in the shop at the time. Then we got orders to come out and inspect the cables on the bridge. So I was up in the top of the tower about three o'clock in the afternoon when we got an after-shock. I'm telling you, it just damn near made me mess my pants. Oh, I mean, the motion was just tremendous up on top. I was getting ready to walk down the cable. I finally did, but it scared the you-know-what out of me.

Seward's father drove rivets and worked inside the towers until a debilitating case of lead poisoning forced him to leave the job and find less demanding work on the San Francisco-Oakland Bay Bridge. On May 27, 1937, the young Dick Seward was taken by his proud father to opening day on the Golden Gate Bridge. Little did the child know that he would return 19 years later to work there. When he talks about the bridge on which he has been employed for nearly three decades, Dick Seward speaks with a pride that is typical of bridge workers.

You can't compare it to other bridges. When you're working on other bridges you're working for a contractor, and like any structural high-rise job downtown, or wherever, you're just working by the hour. You're performing the job they tell you to do and you're going full speed the whole shift. Here, we take our time and what we turn out is beautiful. You couldn't afford to buy it (ironwork) downtown at the price it costs us, but whatever we build for the bridge is built to last forever. You can't do that working for other people.

Seward was introduced to the uncompromising standards which characterize Golden Gate Bridge operations by the man who, at the time, knew more about the bridge than anyone alive, Clifford Paine.

When you work for general contractors and you're out in the field, you put stuff together and you're not very meticulous about it because you've got to get the work done. The name of the game is to make money for the contractor. If the contractor doesn't make money, you don't have a job.

You come out to the Golden Gate Bridge and you'e working for a man like Paine. He had us setting some of the iron with feeler gauges, like in one-thousandths. This is unheard of for ironworkers. If I asked any of my guys out here to take a feeler gauge out in the field to set some iron out there, they'd laugh at me. But that's the way he was.

I got pretty unhappy and a couple of other ironworkers did too. We were about ready to quit. It was terrible having this old man looking right over our necks. He'd come right out on the job and stay there all day long. He's over 70 years old and it's freezing cold out there and you're working out there with a feeler gauge and you have to work without gloves on. We tol-

The bridge is under constant assault from the weather. The Golden Gate draws rain in the winter and fog in the summer. The wind virtually never stops blowing. Passing storms clear the air, their cumulus clouds adding visual drama to an endlessly enthralling scene.

erated him, but that's the kind of guy he was. He was a hard-nosed old engineer who wanted everything meticulous, down to the one-thousandth, and he made a point to stay there and make sure you did it that way. A brilliant man, no question about it. I didn't ever question his ability. It's just that you're not used to working like that.

Seward has seen changes during his years as an ironworker on the Golden Gate Bridge. When he started he was one of three ironworkers; today he supervises a crew of 17. "Work increases as the bridge gets older," he says, and recalls when ironworkers could pick up the bulky paint scaffolds and lug them across the roadway at any time of day. "You wouldn't dare step off the curb now," he laughs.

At one time ironworkers would drive a maintenance truck out on the bridge itself, and park it next to the pedestrian sidewalk in a traffic lane blocked off for their convenience. Now they travel along the walkway itself in three-wheeled Cushman carts similar to those used on golf courses.

Perhaps the ironworkers' most challenging responsibility has nothing to do with wrestling steel. When someone has climbed over the pedestrian rail to one of the girders below the bridge deck and is threatening to jump, bridge officers and highway patrolmen attempt to reason with that individual, hoping to delay the leap until help can arrive. The first person called upon for assistance in such an emergency is usually an ironworker. The tense climb beneath the bridge deck to rescue potential suicides has become a grim addition to the long list of tasks performed by an ironworker on the Golden Gate Bridge.

THE PASSING PARADE

Every weekday at 4:30am, a truck heads across the Golden Gate Bridge. Perched on the back is a worker who methodically stabs round yellow paddles into "slots" in the roadway. The configuration of the six traffic lanes is changed once again: four southbound, two northbound. By 6am the daily commute across the bridge has begun. Nearly 4,000 cars will have crossed by 7am. During the next hour the figure almost doubles, and from 8 to 9am 6,000 more cars will stream across. Between 6 and 10am each morning, Monday through Friday, more than 20,000 vehicles flow south to San Francisco from Marin, Sonoma and Napa couties.

At two o'clock in the afternoon, the paddle truck pulls out of the maintenance yard once more. This time the paddles are replanted to send four lanes north, and the daily commute is repeated in the opposite direction.

For all its awesome qualities, its structural beauty, its unsurpassed setting, there is yet another aspect to the Golden Gate Bridge that sets it apart from other spans: it is a sociable bridge. It welcomes people. A modern bridge, no matter how architecturally stunning, in the end seems to belong to the automobile. But not the Golden Gate Bridge. Despite the thousands of vehicles that cross it daily, it remains accessible to the human experience. Cars, trucks, buses and vans using the six traffic lanes are joined by people on the pedestrian sidewalks who walk, jog, roller skate or bicycle across.

Other bridges may carry more traffic but none accumulates more affection than the Golden Gate. The number of people who visit the bridge every year—not just to drive or walk across it, but just to *see* it—is inestimable. A 1975 poll of travel agents to determine the seven most popular man-made attractions in the United States ranked the Golden Gate Bridge as number one, well ahead of Mount Rushmore, the Statue of Liberty and Hoover Dam.

THE WIND BLEW AND THE BRIDGE SHOOK

The Golden Gate Bridge was 14 years old when, by some accounts, it was almost blown

away. Disbelievers had long been waiting for the bridge to be tested. They expected the day of reckoning to come in the howling fist of a storm, or worse, the convulsions of an earthquake. There had been versions of both during construction.

The public needed some reassurance, especially after November 7, 1940. That was the day the Tacoma Narrows Bridge in Washington — a suspension bridge very similar to the Golden Gate—collapsed in 42-mile-per-hour winds. Chief investigator of the Tacoma Narrows disaster was Clifford Paine. He reassured bridge officials—and, not incidentally, Bay Area motorists—that unlike the Tacoma Narrows Bridge, the Golden Gate Bridge was properly designed to withstand winds of any conceivable force. During the next ten years, the bridge would occasionally buck and sway, exactly as it was designed to do, when attacked by winds of 40 to 60 miles per hour.

The bridge did not, however, behave normally on December 1, 1951. A ferocious storm roared out of the Pacific and through the Gate early in the day. By nightfall the bridge was in the unrelenting grip of winds gusting up to 69 miles per hour. The roadway pitched from side to side in waves. The ripple effect was so drastic that one side of the roadway tilted 11 feet higher than the other. The entire bridge structure seemed to be undergoing violent convulsions. If they continued, it could be fatal.

The ferocity of the storm and its effect on the bridge was vividly described in an official report by chief engineer Russell Cone: "The force of the wind was so strong it was impossible to stand erect on the sidewalk or on the roadway of the bridge. The center of the bridge was deflected between eight and ten feet from its normal position. The suspended structure of the bridge was undulating vertically in a wave-like motion of considerable amplitude. The wave motion appeared to be a running wave, similar to that made by the cracking of a whip."

At approximately 6pm the bridge was closed for the first time in its history. It was a Saturday and toll sergeant Everett Jennings made the decision because higher-ranking officials were not present. Jennings decided the bridge deck was heaving too violently to allow cars across it. Just when it seemed the bridge would not be able to stand up to the pressure much longer, the storm began to retreat. There was an uneasy feeling that nature had pushed the bridge to the limit, then spared it. And although there was no sound evidence to prove it, more than a few knowledgeable observers speculated that if the storm had continued at the same velocity for much longer, the bridge would not have survived.

Even as the winds fell below their 69-mile-per-hour peak, district officials, who had hurried to the bridge when alerted by Sergeant Jennings, continued to be anxious. A cautious inspection of the structure revealed no visible damage. But what if there were unseen wounds that had seriously weakened the bridge? General Manager James Rickets called Clifford Paine at his home in Michigan, described what had happened and deferred to Paine's judgment. Still confident of the bridge's inherent stability, Paine said that if the deck had settled down enough to allow cars to drive in one lane without being blown or tilted out of the lane, it was safe to re-open the bridge. Officials immediately tested it and found they could now drive across in a straight line, something impossible just hours earlier. The bridge was re-opened at 8:45pm.

Paine himself was on the Golden Gate Bridge within a day after the storm. His own inspection uncovered problems, some minor, some more serious. Overall, it was the worst damage the bridge has sustained since completion, and there was cause for concern. Several weeks later, Paine reported to the district board. "For the first time in its 14 years of service, the character, intensity and direction of the wind resulted in movements of the bridge floor of important magnitude. It is believed that much can be done to improve this behavior."

Bridge directors instructed

Paine to conduct an engineering study aimed at reinforcing the bridge against future storms. The result was a series of stiffening girders below the roadway that took nearly two years to install at a cost of $3.5 million—and two lives. In another scaffold accident, a scaffold beneath the bridge deck collapsed and two ironworkers fell 200 feet into the channel. Their bodies were never recovered.

Strengthening it with steel girders after the 1951 storm was the first of only three major repairs needed by the bridge in its nearly half-a-century of heavy-duty use. In the late 1960s, it was discovered that 194 of the 500 steel suspender ropes that connect the deck to the main cables were corroded and needed to be replaced. While they were still able to support the required weight, if the corrosion had been allowed to continue unchecked, they would have been dangerously weakened. Bridge officials decided to replace all 500 ropes, a four-year, $8-million job that was completed in 1976.

The same engineering inspection that uncovered the corroding suspender ropes also revealed serious deterioration of the roadway. By 1972, more than 550 million cars had rolled across the 6,450 feet of suspended concrete pavement. The deck had cracked in numerous places, a normal situation for concrete highways. But this was no normal highway. Into those cracks the fog and winds of the Golden Gate had dumped an ever-increasing deposit of salt. Bridge inspectors had found places where salt and rust build-up had actually raised the level of the roadway on the bridge. Salt content in the concrete was as high as three pounds per cubic yard.

More than 40 years of traffic and weather was wearing out the road, and no amount of repairs could save it. A new roadway was needed. Beginning in 1983, working only at night when traffic is minimal and two lanes can be closed, crews will remove the existing seven-inch-thick roadway and install a new one, piece by piece. The job is scheduled to take approximately two years and cost $50 million, $15 million more than the original cost of the entire bridge! (The 1982 replacement cost of the Golden Gate Bridge is estimated at $450 million.) The road is designed to be replaced in 15-by-50-foot sections of pavement. When completed, some 800 of these sections will have been removed and replaced with a new roadway which is not only stronger and longer-lasting, but actually 11,350 tons lighter than the original.

The bridge approaches the golden anniversary of its opening with a lot of new parts: a new roadway, new suspender ropes, added steel bracing, improved earthquake resistance, even longer-lasting paint. Yet to the millions who arrive every year to visit it, nothing has changed at all. It remains the good old Golden Gate Bridge.

Photo © Mush Emmons

Working only at night when traffic is minimal and two lanes can be closed, crews are removing the old roadway and installing new pavement in 15 by 50-foot sections. The pricetag for the two-year project is $50 million, $15 million more than the original cost of the bridge.